CHILD SPACE
AN ANTHROPOLOGICAL EXPLORATION OF YOUNG PEOPLE'S USE OF SPACE

About the Editor

Dr. Karen Malone is presently Head of Social Sciences in the Faculty of Education at University of Wollongong. She is also Asia-Pacific Director of the global UNESCO-MOST *Growing Up In Cities* (GUIC) project that focuses on children and youth evaluating their quality of life, an external consultant to UNICEF's Child-Friendly Cities (CFC) initiative, and an external fellow for the University Kwala Zulu-Natal. Dr. Malone has published 5 books, 14 book chapters and over 40 refereed publications focusing on children and youth studies, sustainable development, new learning, urbanisation, globalisation, and participatory research methodologies. Her most recent research grant was from the *Australian Financial Markets Foundation for Children* for project where she will be researching with Aboriginal youth around issues of health, well-being and the meaning of place. She is an assistant editor on two major international journals, *Children's Geographies* and *Children, Youth and Environments.* She is the Australian member delegate to the IUCN and an executive member of the Australian UNESCO National Commission social science committee.

CHILD SPACE

AN ANTHROPOLOGICAL EXPLORATION OF YOUNG PEOPLE'S USE OF SPACE

Edited by

Professor Karen Malone

UNESCO Asia-Pacific Director
Growing Up in Cities Project
Social Sciences
Facaulty of Education
University of Wollongong
Wollongong NSW 2522

CONCEPT PUBLISHING COMPANY, NEW DELHI-110059

Cataloging in Publication Data—DK
Courtesy: D.K. Agencies (P) Ltd. <docinfo@dkagencies.com>

Child space : an anthropological exploration of young people's use of space / edited by Karen Malone.
p. cm.
Papers presented at the 15th ICAES Inter-Congress, held at Florence in July 2003.
Includes bibliographical references.
Includes index.
ISBN 818069433X

1. Recreation areas—Congresses. 2. Classroom utilization—Congresses. 3. Children—Recreation—Congresses. 4. Youth—Recreation—Congresses. 5. Child development—Congresses. 6. Educational anthropology—Congresses. I. Malone, Karen.

DDC 305.231 22

ISBN-13: 978-81-8069-433-2
ISBN-10: 81-8069-433-X

First Published 2007

Published and Printed by
Ashok Kumar Mittal
Concept Publishing Company
A/15-16, Commercial Block, Mohan Garden,
New Delhi-110059 (India)

Phones : 25351460, 25351794
Fax : +91-11-25357103
Email : publishing@conceptpub.com

CONTENTS

INTRODUCTION

Karen Malone

Ask any adult what was the most significant space they remember from childhood and many will answer by describing a natural place, away from adult gaze, where they either alone or with a group of friends were able to spend time free to explore and play. Based on a fundamentally idealistic or romanticised view of childhood, we often reminisce of a childhood lost especially in these times of increasing terror, war, poverty and global insecurity. Children around the world, whether in industrialised or developing cities, find they are living in overcrowded, unsafe and polluted environments that provide little opportunity for learning, play or leisure. Dominant discourses abound with the discussions of the forces of economic rationalism and globalisation in late modernity and the impacts these are having on the design and management of unchildfriendly neighbourhoods and communities. Because children are the most vulnerable to environmental and social degradation, in terms of both the likelihood of personal harm and the constraints this places on their capacity to reach their full potential, they are often at the greatest risk of losing out in the battles over the environment. Urban children, in particular, find they are often trapped in environments that provide little opportunity for self-discovery and natural environmental experience. And yet many would support the view that children have a right to recreation, play and education in a safe, healthy and supportive local environment. This has been ordained for years in the United Nations Convention of the Rights of the Child, Article 31 (UNICEF 1992). As signatories to the Convention State parties have a statutory obligation to provide safe, clean and healthy play environment for children within their local community environments. But it is more than a issue of obligation in terms of law it is also a moral obligation that a society and its community have to their children, their future citizens to value and respect their needs and not undervalue the importance of play, recreation and community engagement for children's ongoing health and psychological well-being.

Play and recreational use of space is not only inherently valuable as an enjoyable activity; it is also a process through which children learn. It enhances problem solving and promotes opportunities to support communication, cooperation, creativity, appreciation, responsibility and imagination. Research on children's play around the world has shown regardless of specific cultural contexts, the type, quality and diversity of the place a child lives, has a direct effect on the type, quality and diversity of children's life experiences (Moore, Goltsman and Iacofano 1992) and their potential; for environmental learning (Malone and Tranter 2003).

Local neighbourhoods and villages are important sites for children to develop both social and cognitive skills. The local environment offers a set of *affordances* in relation to the child and their developing sense of self. The *affordance* of an environment is the measure of its capacity to respond and complement the child's identity building. Heft (1988, 1989, 2001) argues that the *affordances* of an environment are those elements it offers or provides for the user and could, therefore, be seen as ecological resources from a functional point of view. They are an objectively specifiable and psychologically meaningful taxonomy of the environment. As the individual's psychological and physical characteristics change, how the resources in the environment are recognised and what they offer also changes for the child. For example, an environment that offers the opportunity for the child to climb or hide underneath elements, or contains features that are manipulative or malleable, is perceived, used and transformed in different ways at different times in the child's life. Behaviour settings are also another way of defining the way a space or place is organised in terms of the child-environment relationship. A distinctive characteristic of a behaviour setting, in comparison to an affordance, is that it is an extra-individual environmental feature; it resides at a higher-level of organization than do affordances. Typically, behaviour settings are not specific to an individual as are affordances because precisely who the individuals are in a behaviour setting rarely matters. The behaviour setting is the way a particular physical environment supports or enhances the potential for the organization of a certain behaviour of individuals with it. For example, a basketball ring whether formally or informally erected in a public space allows for the opportunity for a group of young people to play a basketball – maybe leading to a structured and well-organised community basketball competition. While the setting may exist, it is reliant on the individuals to organise it to be utilised in a certain way. Like affordances, behaviour settings, "...are real entities in the environment, rather than mental constructions. They can be located geographically and temporally (e.g., the soccer game occurred at a specifiable time and place), and their features can be observed and recorded through narrative descriptions" (Chawla and Heft 2002)

Uzzell (1988) also, in relation to the children and their village or neighbourhood environment, argues that there is a psychological benefit to territoriality and its critical role in developing a child's sense of security. According to Uzzell (1998) marking out territories leads to greater identification and feelings of security which can lead to feeling a greater sense of control over the environment with the possible psychological benefits this can confer (1988).

Therefore, we can conclude that the child in terms of place and space has four key child-environment identities. The first is the *physical child* - the child in relation to their movement in and through the physical environment and how that physical environment influences or impacts on their behaviours - this is the affordance of the environment. The second is the *social child* - the child and their growing sense of self and belonging to a community of people, whether it is their family, the village/neighbourhood or the global community. How these relationships have historically and politically come to be and what importance the child places in these relationships is key to understanding such concepts as their growing 'sense of community'. The third is the *natural child* – this identity allows us to explore the child in relation to living things and the ambient environment (i.e. weather, seasons). How the child relates to the natural environment whether it is in a functional sense (for example, growing vegetable for food for family), through play or as part of an experiential learning process. The final is the *learning child* - the child in relation to the environment in terms of their own capacity to utilise the environment for their own learning needs. The child's relationship with the environment as a learning environment is enacted through the spatial, social and the natural environment. By learning how these different dimensions operate together in a dynamic and integrated manner the child then has the potential to develop an identity where they are confident and competent and have the self-esteem and self-determination to function as active independent environmental users, in their present and the future lives.

In this book you are invited to explore the meaning and use of space through the lives of children and young people in a variety of global settings and in the ways in which they are constructing their environmental identities. Often focusing on one or all four characteristics of the environmental identities outlined here, physical, social, natural and learning child, the stories of children's use of space are rich with diversity and contrasts and will not lead to any grandiose generalizations. In fact, in the tradition of anthropological study the shared lens constructed by the authors is not restricted to the physical environment but it focuses on the social relations that become inscribed within these locales, as they exist necessarily *in* space and *across* space (Massey 1994). As Massey makes clear, social relations, networks and the complex politics and power of these relations are played out through the spaces of our shared lives :

> "...as a result of the fact that it is conceptualized as created out of social relations, space is by its very nature full of power and symbolism, a complex web of relations of

domination and subordination, of solidarity and co-operation" (1994: 265).

How these relations are negotiated and how space is constituted through and within them reveals the 'spatial' as being open to the power and politics of 'place' boundaries and the temporal constraints of time. While time is often conveyed as having a "coherence and logic to its telling, while space does not" (Massey 1994: 267), time gives space its fluidity, as de Certeau (1984: 117) articulates so well: "A space exists when one takes into consideration vectors of direction, velocities and time variable. This space is composed of the intersections of mobile elements". The stories presented in this book should then be viewed as fluid both in terms of the space and time that they occupy and the places where they are enacted. 'Places' at this insistence, therefore, becomes shared spaces. They are the sites or contexts where the interests and views of those who feel attached or feel connected to specific locations identify that an area *is*, and/or what it ought to *become* for them across time. Physical locales, therefore through this experience of being in space and time, become 'lived spaces' (Soja 1997), the site for the cultural production of identity (Mathews, Limb and Taylor 2000) and offer the opportunity for young people "to connect the real (materials geographies of place) with the imagined (symbolic geographies of space)" (Mathews *et al.,* 2000: 64).

Additionally, drawing on the work of Bhabha we can see that identity can be produced through and between spaces (Mathews *et al.,* 2000), and that spaces can become, "the terrain for elaborating strategies of selfhood - singular or communal - that initiate new signs of identity, and innovative sites of collaboration, and contestation (Bhabha 1994: 1).

Through its unique relationship with time and place we find that space also has both elements of order and chaos. Order, because all spatial locations of phenomena are caused; they can in principle be explained in 'time'. Order also because there are indeed spatial systems, sets of phenomena in which spatial arrangements are also part of the constitution of a system. And yet, chaos is also an element inherent to our understandings of the spatial, because although the location of each set of phenomena may be directly caused, the spatial positioning of one in relation to the other may not. That is, spatial encounters become unintended consequences - paradoxical mixtures that often end up manifesting as unexpected relations between sets of phenomena. Negotiating and using 'space' is, therefore, by definition complex and ambiguous - seemingly an ordered and simple process while also simultaneously being complex and contested. In no place is this more evident, the nature and purpose of this complex negotiation, then when we gaze into the lives of children and youth. Often positioned as the most susceptible to the positive or negative attributes of a 'physical place' young people's lives embody the multifarious nature of spatial and social networks. What is evident from the chapters of this book is that we can learn much about a place when we engage with children in discussions about their relationships with it. Through these stories and experiences we can see that young people engage in acts of resistance, participation and negotiation within spaces and these acts of spatial interaction are contained within compelling exchanges of spatial compromise between adults and children. For this reason, the construction of how childhood is viewed by adults and children and how this informs the way they are dialectically bound together is at the foundation of the ongoing tussle over space use. These are some of the key themes that are drawn on this book: spatial networks and relations, connections between space, time and a sense of place and the way space users negotiate and construe power in and through social and spatial relations on to places.

When Trawick in the role of anthropologist takes a bike ride into the forest with Menan, in the first chapter of this book, we come to experience 'conflict' and 'hardship' in a challenging landscape, a landscape deeply inscribed with physical, emotional, cultural and political scars. Each encounter on their journey brings a new set of spatial phenomena to be negotiated. Time exists as an ordering thread yet the temporal reality exists only within a cluster of intersections of power and resistance. Trawick's story leads us to look inside and beyond the view of the child as only a 'victim' of war - to be astounded by the strength and resilience of a young man who is actively 'taking up' space while being an active agent of change. The physical child becomes the learning child as we see the extent to which Menan, as the tour guide, 'tells the story of land' and shares that knowledge with the anthropologist.

The next chapter in the collection transports the reader from the rainforests of Sri Lanka to the 'urban jungles' of South America. In her chapter Rabello de Castro discusses the relationship between identity and the appropriation of city spaces through young people's capacity to circulate, to be mobile. She argues for a reconceptualisation of past notions of mobility which often did not delve beyond mapping the ways young people move through 'places' to a deeper analysis where circulation can be understood as a psychological construct that acts to support or impede positions of power and/or privilege.

Continuing with the theme of how power is inscribed onto landscapes by privileging the activities and needs of 'specific' groups and marginalizing the 'other', the next author, Morrow, presents her research with children in UK suburbia. She highlights through the children's experiences the very narrow way 'social capital' has come to be viewed in the literature and why 'geographical' views of place do not adequately identify the complexity of social relations and networks in children's lives. She argues that these social relations exist not in 'place' but in the complexity of heterogeneous 'spaces', often virtual and dislocated. Atencio also focusing in inner city spaces provides a unique look into the role of urban street spaces as contested terrain for engaging in physical activities. He draws on his experiences of interviewing a number of young people between the ages of 13 and 18 in Portland, USA around issues of gender, race, social class, family life and peer relationships. In particular he focuses on the development of gendered identities inscribed in the landscapes and the ways in which masculine discourses position young women as they come to the 'Nike' courts also seeking social mobility through the basketball dream.

Krμger continues to explore the social construction and contestation of space in her chapter on children's use of space in a squatter camp in South Africa. Through the analysis of children's drawings and discussions with the children and their family members she builds on de Certeau's (1984) theory of practice, to expose how hierarchical relationships are exercised through processes of domination and control, resulting in social exclusion. Negotiating these contested spaces means taking risks. What Sibley (1995) refers to as the 'zones of ambiguity in spatial and social categorizations' - the liminal zone, the place where danger and uncertainty lurk for those who are engaging in spatial contravention. Notwithstanding these difficulties, we read of the power of the young people's exploratory urge to 'border cross'.

How childhood space is constructed through a series of socialization processes in the family, in the neighbourhood and the nation, is an extension of these ideas and the focus of the chapter by Kumar. How is a sense of nationhood developed through school curricula and how does this equate to the spaces constructed by adults for children in their daily lives? These questions are explored by Kumar and revisited by Behera when in the following chapter he compares the role of schooling in the life of Tribal children living in Orissa, India. Tribal children, as presented through his intimate observations, find themselves in ambiguous zones - between two spaces, two cultures.

The complex layering of cultural spaces and the ways children negotiate and construct their personal cultural identity, is the focus of the dialogues between the two authors Gold and Gujar with herding children in Rajasthan. The herding children's competence and confidence in being knowledgeable as they go about their daily lives in the grazing lands adjoining the village is juxtaposed and contrasted with their fears and anxiety of being viewed as 'incompetent' in formal schooling settings.

Understanding the relationship between identity and place is also an important feature of the next chapter from Malone. Through a comparative analysis of children living in two different Pacific island settings she exposes the complexity of local-global relations as a social and political construct. Like Gold and Gujar she argues for a view of place identity that encapsulates the coexistence of parallel interrelations to support the notion that children construct multiple and hybrid identities in order to exist in their highly complex local-global, tribal-postmodern world. The idea of a consuming space brings to our attention the role of the voyeur, the gazing other, who invades children's spaces with their own expectations and bias. The child who seems so happy, a blissful simple life on an island paradise - it is too look only at the surface.

The final chapter of the book takes us from the allure of traditional island spaces to the postmodern world of virtual spaces. Not unlike our island children though we find these children are also operating in a parallel universe - the physical and cyber spaces. Pawson, Adams and Tanner bring to our attention, through the transcripts of children engaging in cyber discourses, the powerful impact of social relations and social support. These spaces also provide new ways for children to consider their own identity formation that isn't regulated and controlled through adults intervention, as Pawson *et al.* note in their chapter "Nothing is considered except what they type in and provide access for; they are in total personal control of what they present and to whom."

The aim of this edited collection was to provide a view of space as a practical place (de Certeau, 1984) and yet simultaneously a complex and layered concept that needs further unpacking, particularly when researching and writing with and about children and youth and their use of space. Through these stories of childhood we come to see the very active way children are taking up space: negotiating the politics and differences in space use, being resourceful and competent in its use while all the time conscious of the ways they are often ignored or worse still, marginalized in controlling how space is organized and controlled. The stories here and current sociological research has revealed that although they are frequently positioned as invisible, children are extremely knowledgeable of the multifarious nature of social networks and their role within them. Rather than passive recipients of the environments they find themselves in, children are constantly negotiating and reconstructing spaces in powerful and significant ways, including through acts of resistance. Children's participation in constructing spaces is an important area of research that requires further exploration.

Acknowledgements

A number of the authors of the chapters contained in this edited book were contributors to the thematic session: *Children's Use of Space*, chaired by Jill Kruger for the IUAES Commission on the Anthropology of Children, Youth and Childhood at the 15th ICAES Inter-Congress in Florence. *Children and Young People in a changing World*, July 2003. The IUAES Commission on the Anthropology of Children, Youth and Childhood was initiated after active discussions between anthropologists and other professionals identified a growing presence of sessions, panels and workshops on children and youth at IUAES meetings and conferences. The Commission was officially recognized and its work ratified by the Permanent Council of IUAES on July 19, 2001 during the Inter-Congress that took place in Gottingen. It is through the support of large professional bodies like the IUAES supporting the inclusion of studies on children and youth as important elements of our shared research that the voices of the most silenced and marginalized in our global world will be heard. This book is proof of that fact. As the editor I would also like to personally thank all the contributors for their patience and support during this timely process of publishing the chapters as an edited book.

REFERENCES

Bhabha, H. 1994. *The Location of Culture*. New York: Routledge.

Chawla, L., & Heft, H. 2002. Children's competence and the ecology of communities. *Journal of Environmental Psychology*, 22, 201-216.

De Certeau, M. 1984. *The Practice of Everyday Life*, California, University of California Press.

Heft, H. 1988, 'Affordances of children's environments: A functional approach to environmental description'. *Children's Environments Quarterly* 5: 29-37.

Heft, H. 1989, 'Affordances and the body: An intentional analysis of Gibson's ecological approach to visual perception', *Journal for the Theory of Social Behavior 19*: 1-30.

Heft, H. 2001, *Ecological Psychology in Context: James Gibson, Roger Barker, and the Legacy of William James's Radical Empiricism*, Mahwah, NJ, Lawrence Erlbaum Associates.

Malone, K. & Tranter, P. 2003, 'School grounds as sites for learning', *Environmental Education Research 9*: 283-303.

Massey, D. 1994. *Place, Space and Gender*, London: Polity Press.

Matthews, H., Limb, M. and Taylor, M. 2000. The 'street as thirdspace', in Hollway, S. and G. Valentine (eds) *Children's Geographies: Playing, Living, Learning*, London: Routledge.

Moore, R. & Goltsman, S. and Iacofano, D. 1992. Play for All Guidelines; Planning, Design and Management of Outdoor Play Settings for All Children, MIG, Berkeley CA.

Sibley, D. 1995. *Geographies of Exclusion*, London: Routledge.

Soja, E. 1997. Planning in/for postmodernity, in G. Benko and U. Strohmayer (eds) *Space and Social Theory*, Oxford: Blackwell.

UNICEF. 1992. *Convention on the Rights of the Child.* New York: UNICEF.

Uzzell, D. 1988. *An Environmental Psychological Perspective on Learning Through Landscapes*, Interim report paper 4.

List of Contributors

Matthew Atencio is a researcher with the Physical Education, Sport, and Leisure Studies department at the University of Edinburgh in Scotland. Previously he worked as a project coordinator at the University of Wollongong with an Australian Research Council sponsored grant entitled the "Life Activity Project". His primary interest centres on young people's use of urban spaces and their sport and physical activity cultures. He is also interested in exploring the impacts of "race", class, and gender on youth sport cultures such as skateboarding, street basketball, and hip hop dancing.

Eike Adams is research psychologist at the School of Psychology, University of East London, UK. She has extensive experience working with children and teenagers and her current work focuses on health research using a Foucauldian framework and discourse analytic approach.

Deepak Kumar Behera is Professor of Social Anthropology at Sambalpur University, India. He is the Chairperson of *IUAES Commission on Children, Youth and Childhood.* He is the editor of *Public Images of Children* (1999); *Contemporary Society: Tribal Studies: Volume* 4 (1999); *Contemporary Society: Childhood and Complex Order* (1998); *Children and Childhood in Contemporary Societies* (1998); and *Childhoods in South Asia* (2006). Besides childhood studies, Behera has done intensive research on tribal studies in India.

Lucia Rabello de Castro, is currently Professor of Child and Youth Development at the Institute of Psychology, Universidade Federal do Rio de Janeiro, Brazil. She is also director of the Interdisciplinary Centre for Research and Exchange on Contemporary Child and Youth (NIPIAC) of the same university. She has published in national and international journals and has recently published *Aventura Urbana: Crianças e Jovens no Rio de Janeiro* (The Urban Adventure: Children and Youth in Rio de Janeiro), Rio de Janeiro: 7Letras/Faperj, 2004, as well as edited two other volumes that came out in 2005 *(Mostrando a Real: Um retrato da juventude pobre no Rio de Janeiro* and *Juventude Contemporânea:*

Perspectivas Nacionais e Internacionais). She is Officer-at-large of the Research Committee 53 Sociology of Childhood of the International Association of Sociology.

Ann Grodzins Gold is Professor in the Departments of Religion and Anthropology at Syracuse University and Director of the South Asia Center. Gold's research in North India has focused on religious practices, gender, oral traditions, ecological change and environmental education. Her published four books are: *Fruitful Journeys* (1988); *A Carnival of Parting* (1992); *Listen to the Heron's Words* (co-authored with Gloria Raheja, 1994); and *In the Time of Trees and Sorrows* (co-authored with Bhoju Ram Gujar, 2002).

Bhoju Ram Gujar is headmaster of a Government Middle School in, Rajasthan. He has collaborated on multiple research projects focused on Rajasthan's regional culture and is co-author with Ann Grodzins Gold of *In the Time of Trees and Sorrows: Nature, Power and Memory in Rajasthan* (2002), which in 2004 was awarded the Ananda Kentish Coomaraswamy Book Prize from the Association for Asian Studies.

Jill Kruger is a Research Fellow in the UNISA Department of Anthropology and South African Director of the UNESCO *Growing up in Cities* project. She served on the Codesria Scientific Committee for Child Research in Africa (1999-2001) and is a member for the Commission on Children, Youth and Childhood for IUAES. Her research is ethnographic and participatory and centres on children's sociocultural environments. She has published in various formats and produced professional documentary videos.

Nita Kumar is Brown professorial chair of South Asian History at Claremont McKenna College, California, and Honorary Director of Centre for Postcolonial Education, Varanasi, India. She has published numerous books and articles, both historical and anthropological, on education, urbanism, artisans, modernity and methodology.

Karen Malone is Professor of Social Sciences at University of Wollongong within the Child and Youth Interdisciplinary Research Centre and has since 1996 held the position as the Asia-Pacific Director of the UNESCO *Growing Up in Cities* project. She has authored numerous edited books, chapters and journal articles in the areas of children's environments, rights, identity and participation. Her most recent research grants are from the *Australian Financial Markets Foundation for Children* for project where she will be researching with Aboriginal youth around issues of health, well-being and the meaning of place and *The Smith Family* where she will be focusing on how to design child-friendly cities with children identified as 'at risk'. She is an assistant editor on the two major children's environments international journals, *Children's Geographies* and *Children, Youth and Environments* and has recently edited 3 special editions of international journals. She is the Australian member delegate to the IUCN and an executive member of the Australian UNESCO National Commission social science committee.

Virginia Morrow is Lecturer and Course organiser of MA Childhood Studies (based on the sociology of childhood and children's rights) at the Institute of Education, University of London. Previously she worked at Brunei University, London School of Economics (Gender Institute and Sociology Department), and the Centre for Family Research, University of Cambridge. Children and young people have been the primary focus of her research activities since 1988. Her research interests include qualitative research methods and the ethics of social research with children; history and sociology of childhood; social capital; children's work; children's understandings of family, children's and young people's perspectives on their environments; children's rights.

Chris J. Pawson, School of Psychology, University of East London, UK. Chris is a developmental psychologist and lecturer at the University of East London. His major research interests include children's social knowledge, peer relation's and the development of pro-social and antisocial behaviour. Chris is a founder member of the Commission for Children, Youth and Childhood (International Union of Anthropological and Ethnographic Studies).

Ralph Tanner has taught Sociology and Anthropology in several Universities in East Africa and London and is a retired Chairman of the East African Institute of Social Research in Uganda. Dr. Tanner has published widely on crime, religion and childhood, as well as research methodology, drawing on over 40 years of fieldwork in Africa, Asia and the United Kingdom.

Margaret Trawick is Professor of Social Anthropology, College of Humanities and Social Sciences, Massey University, Palmerston North, New Zealand. Her research and writing is about the Tamil-speaking world. Her publications discuss South Asian medical systems, verbal art both oral and written, the cultural organization of emotion, relations between male and female, and warfare. Her books include *Notes on Love in a Tamil Family* (1990) and *Enemy Lines: Warfare, Childhood and Play in Batticaloa* (2007).

Chapter 1

FREEDOM TO MOVE

A Bike Trip with Menan

Margaret Trawick

In 1997 and 1998, I went to Sri Lanka to do ethnographic research on children who had been affected by the armed conflict there.[1] I was interested in how children might exercise agency in such circumstances. I did not have a clue what I would find.

From the point of view of many westerners, the involvement of children in warfare is most dramatically horrific because of the physical and psychological damage caused to children by the experience of violence. Even if a child is not herself physically wounded, merely to witness a moment of severe violence is thought, by western psychologists and those imbued with western psychological thought, to be traumatic for children, as well as for adults, and to leave lasting internal scars. There is a view that exposure to violence can result in the mimetic infliction of violence upon others, and in this way a vicious cycle of violence may be formed. The psychological damage of exposure to violence is considered to be greater for children (including youth) than for adults.

But I learned during my fieldwork that the most pressing hardship for young people in the war zone of eastern Sri Lanka was not the witnessing of violence, no matter how gory. Even having violence inflicted upon one's own body was not in itself the greatest hardship. Rather the greatest hardships felt by the young were certain specific consequences ensuing upon active warfare and affecting civilians as well as combatants - in particular, separation from family, loss of loved ones, and loss of control over personal space and movement. These losses were often, but not always, concurrent.

Batticaloa is the name of a town in eastern Sri Lanka where I worked, and it is also the name of the district in which the town is situated. The population in the region is almost entirely Tamil. For more than two decades, Sri Lanka has been torn by war between the Government of Sri Lanka and the Liberation Tigers of Tamil Eelam (LTTE) who are fighting to achieve independence of Tamil people and the contiguous areas they have traditionally called home from Sri Lankan Government control. The majority population of the island of Sri Lanka is Sinhala. Tamil people are the largest minority. Sinhalese and Tamils speak different languages, and have different histories and mythologies. The armed conflict in Sri Lanka is one of the many so-called "ethnic" conflicts raging throughout the world.

This war, like many others, is in large measure about control of territory. The Tigers and the Army have fought mile-by-mile and village-by-village to gain and establish control over bits of land and the people in them. Some towns and villages are continuously fought over, with both a heavy Army presence and a heavy, if covert, LTTE presence. In the region of Batticaloa, the situation since 1996 has been that the Army controls the coastal cities and the Tigers control the areas inland. The coast is separated from the inland by a set of long, narrow lagoons. A lagoon makes a natural boundary between territories, and to cross the lagoon is to move suddenly and sharply from one world to another. One may cross a lagoon by bridge, by ferry, or by boat. One may also swim across, although crocodiles present a danger.

All the traditionally Tamil-owned areas are contested, and it is in these areas that the war is fought. Civilians have to live their lives in the midst of the fighting. In the Batticaloa area, some stay entirely on one side of the lagoon or the other. But others must cross almost daily to carry produce or firewood from the hinterlands into the towns to be sold, to bring provisions from the towns into the hinterland, to visit with relatives, to travel from home to work.

The crossing is always dangerous, because one must pass through checkpoints, where one may be detained, beaten, taken captive or killed by the checkpoint guards, or where one may be subject to having belongings confiscated. However, staying at home is no sure protection either, because raids into villages and homes by police, army, and paramilitaries are not uncommon. In addition, until the formal ceasefire that began in February 2002, bomb explosions in public places such as markets or commercial buses were frequent, and shelling of villages from distant ground installations, as well as from airplanes and helicopters, happened rather often as well. During intense phases of the conflict, people have fled from one village to the next, from house to school or temple, from inhabited area to jungle, from countryside to city, and from region to region throughout the island seeking safety. Those who can leave Sri Lanka altogether have done so. But many do not have the facility to leave.

Youth, especially young men, are the most likely to be targets of violence or capture by Government forces, because they are of an age when they could be active fighters in the LTTE. Women are not exempt from suspicion, as the LTTE has women fighters also. And a civilian of any age, even if unable to fight, can still be a Tiger "helper". Still, it is mainly young men and adolescent boys who are suspected, and therefore detained, interrogated, and worse.

Adolescence is by nature a time of increased exploration of the world outside of home, and also a time when personal self-determination is uniquely important. Restriction of personal mobility is difficult for anybody. So is forced mobility. But in Batticaloa, youth are exactly the ones upon whom both forced mobility and forced immobility are most likely to be imposed, and they are the ones most likely to chafe against this loss of personal physical freedom.

It is easy to obtain a list of the specific problems presented by the war to civilians - one need only ask them. In my interviews and conversations with youth and adults they discussed their most serious problems. All of these problems entailed loss of control over personal mobility.

Checkpoints were the most frequently mentioned, and in some ways the worst of the problems young people experienced in connection with the war. Every person who went through a checkpoint was subject to bodily search by armed guards. At the minimum, these checkpoints slowed travel.

Displacement was not an everyday occurrence, but it was probably the most terrifying experience in the children's lives. People who had stayed this long (i.e., through all the horrors up until 1998) would only flee their homes if they were in immediate danger of attack by the Army – i.e. if they could see the Army members coming with combat rifles firing, or if bombs and mortars were falling very close. People who fled would sometimes stay in the jungle, where drinking water, food and shelter were scarce, and poisonous snakes and other dangerous creatures abounded. Wherever they went they had to hide. But the worst thing of all was that in the chaos of flight, family members could get lost from each other. For a young person, or anyone, to be lost and alone, even for a few hours or days, could be enough to disable all natural courage in the person for years or even a lifetime.

Abduction also threatened these youth. Little children and older people would not usually be abducted from their homes or at checkpoints and taken prisoner, but youth would be. Paramilitary groups would come at night to houses on border areas and capture young people from their homes. The abducted would be imprisoned and interrogated, and many of them would never be found.

Being trapped in a round-up was another source of fear to the young whom I spoke with. Government forces would surround a whole neighbourhood, supposedly looking for Tigers, and would close in around, leaving no room for escape.

Landmines did not choose their victims; a landmine would explode when anybody set foot on it. For this reason, nobody could walk without fear through certain areas – but young people were the most likely to be careless. Landmines thus were another restriction on normal youthful mobility. Moreover, if you did happen to set foot on a landmine, if you were not killed, you would lose pieces of one or both legs and subsequently would be unable to get around easily. It was not the disfigurement, nor even the pain, that caused most distress to the young landmine victims I knew, but the fact that they could not bicycle or run swiftly to school or anywhere anymore. Other children would laugh at them and school teachers would scold them for being late. (We cannot expect that school children and school teachers will always be compassionate).

Finally, the choice whether or not to join the LTTE involved issues of mobility. Tamil youth, and youth only, were targets of Tiger recruitment efforts, and there was a certain fear among some that they would be taken unwillingly from their houses to a Tiger training camp far away, there to be trained and sent into combat. Even youth who joined voluntarily would be taken away from their families and not allowed to see family members until they (the youth) were securely committed to remaining with the LTTE. On the other hand, membership in the LTTE entailed, for some youth, a degree of mobility, freedom, and even safety that they did not have in civilian life. They got bicycles and motorcycles and could go where they pleased after they were ensconced in the movement. For a young Tiger girl, if she went about in public in uniform, no male or female civilian would dare harass her. The Tigers were protected by their rifles and other armaments, as well as by the organized intelligence of the movement, which enabled them to know in advance (sometimes) if the Army was coming, so that they could get away.

The following narrative describes some experiences and conversations I shared with Menan, a sixteen-year-old boy who lived and still lives in Batticaloa. Like many young people in the community, Menan understood and accepted the cause of the LTTE. But he did not want to join them, knowing that his life would be further jeopardized if he did so.[2] He was his mother's only child, and he needed to care for her.

At sixteen, Menan was on the cusp between childhood and the dangerous age of deliberate entrapment by one or the other side. There was also a certain danger that he might become entrapped by me. No matter how carefully I tried to explain my agenda and show myself trustworthy, Menan could not know for sure that I was not working for one or the other of the warring parties. But he did know that, as a sympathetic foreigner, I might be in a position to help him out in some way in the future, although I never made any promises in this regard. Maybe he even really liked me and cared for me - not as a mother surrogate, for his own mother was with him and loved him above all others - but as a fun and not-too-vigilant Auntie. For whatever reason, he did not avoid me, but rather sought me out and found me at times of his own choosing. He kept himself in control of our interactions, catching me off-guard many times, but never letting himself be caught. At the age of sixteen, Menan had almost complete autonomy. His mother and other relatives, despite their concern for his life, allowed him to negotiate the perils of the world on his own. Perhaps this is among the reasons for Menan's apparent ease in the face of hardship. As Boyden (2003: 8) identifies: "...there is some evidence that children who try actively to overcome adversity - by attempting to resolve the problems they face, regulate their emotions, protect their self-esteem and manage their social interactions - are likely to be more resilient than children who accept their fate passively, especially in the long run" (see also Cairns, 1996; Beristain *et al.*, 1996; Garbarino, 1999). The fact that Menan had a strong safety net in the form of his kin and their land and resources also probably lent him confidence.

I learned from Menan that rather than attempting to "trap" him in verbal interviews, I was able to understand more about him by travelling with him where he wanted to take me, and by letting him borrow my paper and paint, when he asked, and letting him take these materials away and bring back the finished products when he decided to do so. Coles (1986) and Das (1991), among others, have written extensively about certain children's communication of their thoughts through drawings and bodily demonstrations rather than through words. Most especially, when the children are in danger or in trouble, communication in this way is crucial. Both Das and Coles also showed how such children exercised agency (changing their worlds and making the people around them pay attention to the children and do what the children wanted) through bodily action more than through words. Both drawing and bodily action, unlike spoken words, involve the use of space. Via these two devices, he showed me the world through his eyes.

My narrative is a clip of field notes I took about Menan in 1998. He understood that I was writing about him, and that he could ask me to stop writing about him whenever he chose. I told him often, and he understood, that he was free to talk with me or not. But, as I have said above, he often sought me out. I gave him my notes about him to edit and comment on before I made them known to others. He seems to have been, and still to be, happy for the recognition.

Menan and Me

Menan is an only child. The Sri Lankan Army killed his father in a massacre of civilians in 1987. Menan was four years old at the time. I first met him in January 1998, when I was staying in the Batticaloa area of eastern Sri Lanka for a long stint of fieldwork. He was fifteen years old then but looked to my eyes around ten. He lived with his mother in the town of Batticaloa where he attended school and was first in his class. When I told my friend Vasanta about the nature of my research, she said I should meet with Menan, whose father was Vasanta's brother. Because of the way his father died, Menan was said to be a war-affected (*paatikkappadda*) child. In fact, there was hardly a person in the Batticaloa area who was not war-affected, but if you lost a close kinsman to the violence of the war, you were considered more war-affected than those who had not.

It is January 1998. Vasanta says of Menan that he prefers to be on this side of the lagoon rather than in Batticaloa with his mother because he has more freedom over here. His mother is always telling him to watch out for the paramilitaries. Also there he has school; here he stays during holidays. On the other side are curfews and checkpoints. On this side Menan can come and go as he pleases.

Today is the first Saturday of the New Year. Menan turned up yesterday and we decided on the spot that we would take the bike trip into the jungle that we had planned before but never taken. I ask Menan where exactly we will go. We are going to a nearby low mountain called Thaanthaa Malai on the top of which is a Murugan Temple. It is somewhat dangerous to go there because an LTTE jungle camp is somewhere out there. As long as we do not stumble upon that camp, we will be okay. But if we do happen to discover it, the people there may kill us.

Today it is cloudy when we start out but soon it begins to rain, and the rain falls heavily all through the day. It is the rainy season so this is to be expected. The road is terrible - sandy and muddy and wet and full of holes and puddles and rivulets. We cycle through the rain over the muddy road - it is like playing the game of labyrinth, you have to steer your bike with two steady hands through this obstacle course of potholes and mud puddles and narrow slippery bunds. Pedalling through mud is like pedalling uphill except there is no downhill when you come back. Menan does it with ease. I follow close behind him and only fall down twice, landing both times in the foot-deep water of paddy fields. Menan shames me by telling everybody we meet that I have fallen off of my bike. It doesn't help that the brakes on my bike don't work, but probably neither do Menan's. At first we encounter quite a few people biking the other way. They laugh or smile when they see us; some of them ask Menan where we are going and who I am. The two of us must make a funny sight.

Before long the traffic thins out and Menan and I are travelling all alone through the rain on this awful road. The landscape is beautiful - there are paddy fields as far as the eye can see, interspersed with groves of palm and bits of thick wild vegetation.

After we have been biking for about an hour or so, Menan turns off the road onto a path that leads through the paddy to a palm grove where there is a tiny hut. I follow him, not knowing why we have turned off the road, but guessing that Menan has decided to seek shelter because it is raining so hard. We reach the edge of the clearing and Menan calls for someone, but there is no answer. The hut is empty. Menan beckons me to follow him on a path that leads from that palm grove to another palm grove with another hut. We go there and that hut too is empty. Menan says we have to go on down the next path to the next grove. It is still raining heavily and we are about a mile off the road in the middle of paddy fields and now I am really confused. But Menan says that definitely there will be someone at the next place to which we are going. So with a sigh I follow.

We come to a larger palm grove with several structures in it, made of sticks and palm leaves and mud and thatch. And two men are there. They greet us. "This is our land," says Menan. And now I understand why we have taken this detour. He wants to show me his family's land: his land.

One of the two men is Menan's grandfather, whom Menan calls *appappaa*. This is actually Menan's mother's father. The other man is a neighbour. I met Menan's grandfather once before, in town, where he had come to fetch a vet for his goat that was having trouble giving birth. The next time I saw Menan's grandfather I asked him how the goat was and he said it had recovered. Now there is a pair of cute twin kids in the palm grove, but Menan's grandfather says the mother goat has died. Like all homesteads I have seen in the area, this one is cleared and has a floor of smooth raked sand and is fenced all around. The fence encompasses a square about seventy feet by seventy feet; the edge between the clearing and the fields is distinct. It is an island in a sea of paddy. The tall palm trees make it look even more like a fantasy island, and so do the thatched huts. The scene resembles the pictures of natural landscapes that children and adults often make for me using my paper and paints.

One of the thatched huts has a sleeping room and a covered porch in which is room to sit. A separate hut is for cooking. Both of those huts have mud walls. The thatch is carefully laid and tied. Stored in the veranda of the sleeping hut is a very rusty bicycle with tires that have not entirely lost their tread. In a corner across the centre of the clearing from the sleeping and cooking huts is a goat shed made of sticks with a thatched roof. The shed is raised above the ground like a rabbit's hutch. Next to the goat shed a vine is growing on a stick trellis with gourds hanging down from it. Each gourd has a quarter of a coconut shell tied to its end - so that it will grow straight and not curl around, the grandfather explains when I ask. In addition to the tall palms there are smaller younger palm trees and some fruit trees. The whole place is immaculately kept. The only trash I am able to discover lies in a hollow at the base of one of the palm trees. This trash in the hollow consists of one broken glass jar and a pair of old sneakers. I am not sure whether even these things are trash in the sense of having no use.

Both Menan and I are completely soaked. Menan goes into the sleeping hut and changes out of his wet clothes into an old piece of sari (his grandmother's) which he wraps around his waist to cover the lower half of his body. I ask for a sarong and am given one which I wear while my jeans attempt to dry.

In the sleeping hut we eat the lunch we have brought, and, the rain having stopped, Menan invites me outside. He does not explain what we are going to do; I just follow him. As we go along he points out every plant and names it for me. He picks a bean for me to eat from a bean plant. He picks a small flower and hands it to me, and later a big flower. "For a bouquet," he says with a smile. We come to a large boulder and he climbs up to the top of it and tells me to come up, too. I am tempted at this point to remind him of the age differential between us and certain consequent differences in agility, but I stop myself and simply say to him as I face the steep wet boulder, "I'm not sure I can."

"Try (*try pannunka*)," he says tersely. Obediently, I do make an effort to climb the boulder and shortly am at the top. There is a *peepal* tree growing out of the boulder and a neem tree growing in the shade of the *peepal*. I ask Menan whether anyone planted them that way and he says no, they just grew.

The view from the top of the boulder is lovely, and so is the homestead, and so is the landscape as we walk through it with Menan pointing out all the different kinds of wild and domestic plants that grow in close company with one another on his family's land. I tell Menan how beautiful I find the whole place, and he affirms, "Yes, this is nature (*iyatkai*)." As we walk back we come to a tallish lemon tree and Menan tries to beat a big green lemon down from the top as I wince for the tree. The big green lemon hangs on fiercely but smaller ones fall and I gather them up. Around noon we decide to resume our journey and we change back into our wet clothes. The rain has also started again. There is a short-cut through the paddy fields to the mountain temple we are aiming at but Menan says he doesn't know the way; he will have to get someone to show us. A man is wakened from his sleep in a hut in a nearby homestead and this man leads us riding Menan's bike while I ride my bike and Menan rides the old rusty one from his grandfather's veranda.

We make our way over ditches and bunds to the rocky hill where the temple stands. It has just been built about twenty years ago. The climb to the top is easy. Because of the rain, we cannot see far, but are told that an army camp is not far away from here, behind a hill. It is confusing to me, because I thought all the Army camps in the hinterlands had been abandoned two years ago when the Tigers took over the area, occupying the abandoned camps. But Menan and the man with us insist it is an Army camp. At the base of the temple hill is an *ashram* where men and women are staying. Some, maybe, all of them are *sannyasis*. They feed us a big meal of a rice dish and fruit and insist that we spend the night there with them. There is no point in going all the way back through the rain, wait until it stops, they say.

We are in what people call the jungle, but surrounding us are all boulders and scrub. A scraggly peacock appears, and some monkeys. It is chilly and wet in the ashram, and we have no dry clothes, so we decide not to stay the night, but to return home today. All the while I have been worrying aloud that Menan will catch cold from this trip. If he does, his mother will kill me. He is a small, frail-seeming child and respiratory infections are a cause of serious suffering and danger to people here. For me, the heat is the problem, and the rain is a welcome relief. I could not have made this trip on a sunny day. But Menan insists that he will be fine.

When we get back to Menan's grandfather's hut, the rain has stopped. He too insists that we stay the night and return in the morning. He gives us fresh milk from his cow. It is the only fresh milk I get during my whole time in Sri Lanka. But we decide to return to the village today, so that people won't worry about us. Menan's grandfather accompanies us across the fields towards the main road. Menan has asked his grandfather for a cuckoo's nest. And so his grandfather takes us to a hedgerow where an abandoned cuckoo's nest is hanging from a tree branch. It is like an oriole's nest - a kind of stiff gourd-shaped bag that the bird has fashioned from straw and glue, with a hole in the side for the bird to go in. Because it is pretty, because it is a small marvel of avian craftsmanship, it is valued, and we take it home.

On our way towards home, the rain begins again. Menan asks me if I like bicycling in the rain. He says he likes it a lot. I say it is a new experience for me, and that's why I'm not very good at it. He says it is a new experience for him, too, biking over this kind of muddy road in the pouring rain. And that's why he likes it. The people who live here do it everyday, he adds.

I interview Menan the day after the bike ride. A good taped interview for me is like gold - it can throw great light on people's thoughts, worlds, language, everything; but only of course if they want to be understood, which not everyone does. I am hoping that Menan will just ramble on about his thoughts, as many folks do, but perhaps because of the formality of the situation, most of Menan's answers to my questions are brief. In general, Menan is not a talkative boy. And of course he has to be careful what he says. Still, he says some interesting things.

Menan tells me his favourite movie is (*Athisiya Piravi* or "Miraculous Birth"). His favourite story is a Muslim tale entitled *Niithikkup pin Paasam* ("Justice Before Love"). In the interview, Menan tells me the plots of both.

The movie is about a boy who dies prematurely and has to go back to earth, but when he gets back his body has already been burned so he must choose another body. The person whose body he is forced to inhabit has one girl friend, while the boy who inhabits that body has another girl friend. He is not allowed to tell the truth of his situation, or his head will explode. Thus, the movie places its hero in an impossible situation, not of his making. He does what he can to escape having to tell the truth, but finally, he is forced to do so. Then the highest god intervenes and saves him and everyone lives happily ever after.

The Muslim story allows a queen, who has accidentally killed a commoner, to compensate for the murder with money, and to buy her own life in the same act. The story is ostensibly about equality of people before the law, regardless of who their kin may be. But the story also takes for granted that life has a cash value, and this trading of money for life is an event so common for civilians in this area that its fairness is not even questioned – at least not by Menan. His father is dead, but he will make out because he has land and is getting a good education.

Both the movie and the story reflect to a certain extent the situation of Tamil people caught between two armies. If they tell the truth, they may be killed. And they find themselves, together with Muslims, to be the victims of ethnically biased "justice".

Menan makes it clear that he has no intention of joining the LTTE. He does not want to fight. But he thinks that what the LTTE say (about what they are fighting for) is "correct". In subsequent communications, which I will discuss below, he conveys to me his ideological view in detail. In general, it is in accord with LTTE ideology, except that Menan does not believe that fighting will solve any problems. He considers the fighting itself to be the problem. His is the pacifist's dilemma. He wants the fighting to end, but he says that "we" can do nothing about it, only "they" (the people themselves who are fighting) can end it.

In Menan's view, his greatest problem is absence of personal freedom - in particular, the obstacles such as checkpoints that restrict his mobility. In general (like most adolescents?) he seeks "freedom from control by others".

When I ask Menan, "What would a truly happy life be like?" he responds, "I could go anywhere at any time, I could see anything without restriction by others. I could travel freely anywhere at night. I could go anywhere that I wanted to go." And the places he wants to visit are exactly those that are forbidden to him: the old fortress in town, where the Army is now stationed, and another area where there are only farms, but where he cannot go because the Army thinks the LTTE are there. These restrictions bother him more than the bomb blasts and fighting that are happening all around. Indeed, it seems that the bomb blasts and fighting (the "problems" or *piraccanai*, which consist precisely of war-making) are troublesome to him because they are what restrict his freedom. If the fighting stops, he believes he will have the freedom he wants. The concept of fighting for freedom is not brought up in the interview.

As a small child, Menan did not fear the Army, because he believed that they would not hurt small children. Now he fears them because he is getting bigger, and there is a danger that he might be pulled into the conflict. And he fears the bombing, as anyone would, because no one can tell when or where the next bomb will fall or explode. But his experience with such things seems to mute his fear, rather than intensify it, and any fears he may have are overridden by his great desire to explore his world.

Menan says that he is accustomed to the events of war - in his words he has experience (*anupavam*) with such events. For Menan the events of warfare - bombs exploding and so forth - are customary (*vazakkam*). Therefore, he has no interest (*aarvam*) in them. Adults, he says, have no (previous) experience of the war, and therefore, it is interesting to them and therefore, they talk about it. Menan says he and his friends do not talk among themselves about the war.

Menan's analysis of the difference between his attitude and that of his elders contrasts with what his elders might say. Menan has grown up in the midst of warfare, and that is what he means when he says he has "experience". His elders grew up in peacetime, and when war came down upon them, it was something new to them. For Menan, new things are exciting and interesting, old things are dull and boring, and he surmises that this is why adults are so obsessed with the war. In a sense I think Menan may be right. Warfare is exciting but only because it is so different from what one perceives as ordinary. In another sense, however, Menan does not have experience with the war. He did not witness the death of his father, and was too young at the time to fully understand its importance. On the basis of his interview, one might guess that the horror and outrage have never really come home to him.

Menan has experienced no serious problems with the Army, no problems at all with Sinhala civilians. He has never met a Sinhala civilian except on the school trip he mentions. The reasons for the war are abstractions to him. He wants nothing to do with the war, he does not even want to talk about it, unlike older people who talk about it all the time.

In addition to finding the war a boring topic of conversation, Menan finds it depressing. He states that the most important event in his life was the death of his father, but he does not even want to talk about that. Talking about his father makes him sad, and he would rather do things that make him happy. What gives him happiness is making jokes with his friends and laughing.

Menan does think about the war, as he makes clear to me later. But in this interview he says he does not think about it "much". He makes no reference to the war when he talks about his future, about the things he loves and the things he wants to do. He clearly loves his family farm and when he is on it he strides over it like a king. He does not imagine losing it or abandoning it, and he says he would be "happy" with farming, if that were what he had to do. But his future is in learning and in teaching. Education is in large part for the purpose of getting a job, and provides a "future" in that sense. But there is more to it than just that. For Menan, who is first in his class, and thus a very good student, learning is becoming and growing and discovering the world. So he wants to be a teacher - anywhere in Sri Lanka, and not necessarily in his home town. He wants to share his experience with small children, and thereby make them happy. When he makes this statement, he must not be thinking of the fact that to date his experience is of a war-torn world without freedom. Menan's love of the family farm with all its growing things, and his intention to teach small children and make them happy seem to me part of the ethos of regeneration that pervades this whole area, war or no war.

The interview with Menan was my idea. I asked questions and

Menan answered. But the bike ride was Menan's idea. He led and I followed. Inasmuch as I was there to understand children's agency, the contrast between the bike ride and the interview taught me an important lesson: if you want to know how children or youth exercise agency, let them do it in their interactions with you. Let them take the initiative and let them make the decisions. They will understand and decide pretty quickly what their relationship with you is to be.

In addition to this basic lesson, my time with Menan reminded me of how some people prefer to demonstrate their thoughts to us rather than putting them into so many directly spoken words. During the bike trip few words were exchanged, but a beautiful world was shown to me.

I have brought over to Vasanta's house some paper and crayons and water paints, so that children can draw pictures for me. One day Menan asks to have some paper and paint, and I give them both to him. Days later he comes back to me with a picture he has made.

The landscape is diagonally divided by a straight stretch of blue water. On the upper left half of the landscape grow tall and perfect trees, and a man is planting another tree there. The bank of the water on that side is clear and clean. The opposite bank of the water is littered with dead tree branches, and the trees in the lower right half of the landscape are broken and losing their leaves. A man on that side is chopping a tree down.

Menan gives me an essay he has written to accompany the picture. Translated from Tamil, the essay reads:

> "In the world today, trees are being cut down everywhere. Therefore drought and famine are arising everywhere. The resources (*vaLankaL*) of Lanka are being affected (*paatikkapadukinratu*). Therefore the environment is affected. The environment is polluted (*maacupadukinratu*). Because trees are cut down, the light of the sun dries up the waterholes directly, and drought occurs. Therefore today, people and animals are dying for lack of drinking water.

Because trees are cut down, the weather becomes disorderly (*vaanilai kuzampukiratu*). In the air, the CO_2 increases. Therefore, the heat in the environment increases. When trees are cut down, in the rare world some species of creatures that grow only in a few places and to a limited degree and rare plants die. Species of creatures that grow in the environment nearby are affected. Even species of creatures that live here are dying.

Beyond the river a young boy is planting a tree. Therefore that environment grows with great flourishing. Therefore on that side all the species of creatures live freely (*cutantiramaaka*). There the natural resources are unaffected. It presents a very flourishing sight. We too should follow that boy's example. We should protect our natural resources.

One may find many connections between what is drawn here in this picture and our country of today.

Here the destruction of trees happens in the north and in the east. In other words, the Lankan government has entered into our region and destroys our species (*ematu inam* – the term in modern usage means "our people," that is, the Tamil people). Our freedom and our rights are being cut down. Our life's order has been mashed (*ciir kuzaintatu*). The Lankan government constrains, suppresses, and destroys the Tamils. It wastes our future. It destroys our relatives. If we are beaten so, why should we stay alive?

As for the other side, the Lankan government makes the region flourish where the Sinhala people (*singala in a makkal*) live. And it bestows upon those people all the rights that are needed. It uplifts the lives of those people. It protects only that category (*inam*). This is what is happening in Lanka."

If one looks at this picture, one will see the government that runs Lanka. Calling itself a democratic, people's government, it constrains one category of people and plants concessions (*aravaNaittu naduttukinratu*) for another category of people. In the same way, here the trees on one side are cut down. The trees on the other side are protected. The Lankan government protects one side, that is the Sinhala people, and destroys the other side, the Tamil people.

Menan signs the essay in English with his full name, age, school, and home village. I think the essay needs no commentary but for the obvious observation that Menan expresses his thoughts to me more boldly and clearly in writing and painting than in speech.

A week or so later, Menan brings me another painting. A blood-red sea roils against a black sky. Pale human hands reach out from the sea, like drowning people's hands. And scattered across the sky, in bold white English capital letters, are the words, "HELP! HELP! HELP!"

The message could hardly be clearer. But Menan has written a few sentences to explain the picture. He writes:

> "Here are the hands of Tamils drowning in blood. Why? Because of the cruelties of the Lankan government. Singing and singing refrains that say, "Peace, peace," the Lankan government destroys poor innocent Tamils.
> Because of the Army's cruelties that the Tamil, struggling for life in a sea of blood, cannot endure. Today the Tamil asks the world's nations for help. Hands of the world! Will you not change into helping hands for us?"

And later still when I am soon to leave the country, Menan brings me another painting. It is a night scene, therefore, there are no colours but white, black and grey. The river, straight and calm, stretches from the foreground to disappear between black distant mountains. The fullmoon shines above the mountains and gleams upon the river. Two dark figures, one larger and one smaller, are crossing the river in a boat together (Fig. 1.1).

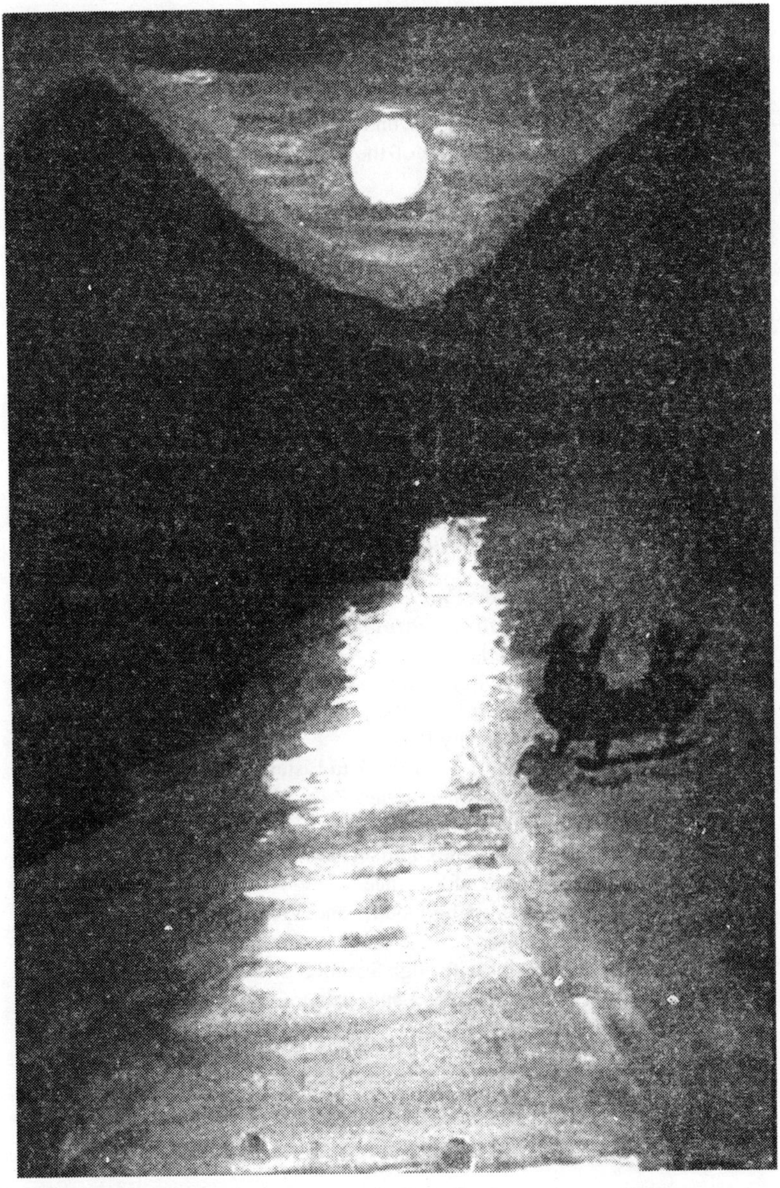

Fig. 1.1

NOTES

1. The research on which this paper is based was carried out in Sri Lanka from November 1997 through June 1998, under a grant from the American Institute of Sri Lankan Studies. The title of the research project was "Children as Sociocultural Agents in Sri Lanka."
   In a major review article published just last year, the comment was made: "Children's own voices and perspectives have been largely absent from the anthropological literature on childhood and violence" (Korbin 2003: 431). I have attempted in my own research to help correct this problem.
2. For more about Tamil youth's experiences in and views concerning the conflict in Sri Lanka, see Trawick 2002.

REFERENCES

Ahearn, L. 2001. "Language and Agency", *Annual Review of Anthropology*, 30: 109–37.

Austin, J.L. 1962. *How to Do Things with Words*. London: Oxford University Press.

Boyden, J. 2003. "Children under Fire: Challenging Assumptions about Children's Resilience", *Children, Youth and Environments* 13(1). Retrieved June 1, 2004, from http://cye.colorado.edu

Coles, R. 1986. *The Political Life of Children*. Boston: Houghton Mifflin Company.

Daniel, E.V. 1996. *Charred Lullabies: Chapters in an Anthropology of Violence*. Princeton: Princeton University Press.

Das, V. 1991. "Our Work to Cry: Your Work to Listen", (pp. 345-399) in V. Das (ed.), *Mirrors of Violence: Communities, Riots, and Survivors in South Asia*. Delhi: Oxford University Press.

Das, V. and Reynolds, P. 2003. "The Child on the Wing: Children Negotiating the Everyday in the Geography of Violence."

Korbin, J.E. 2003. "Children, Childhoods, and Violence", *Annual Review of Anthropology*, 32: 431-446.

Trawick, M. 2002. "Interviews with High School Students in Eastern Sri Lanka", (pp. 366-380) in D. Mines and S. Lamb (eds.), *Everyday Life in South Asia*. Indiana: Indiana University Press.

Chapter 2

CIRCULATING IN THE CITY

From Routes to Projects, From Diversion to Conversion[1]

Lucia Rabello De Castro

> *"Knowing how to orient oneself in a city doesn't mean much. However, getting lost in a city, as if you were lost in the woods, requires direction."*
>
> *('Tiergarten', Childhood in Berlin, Walter Benjamin)*

Circulating in the city is imposed on us as a condition of urban dwelling. In order to visit friends, go shopping, study, have fun, among other activities that young people are used to do one has to penetrate the maze of streets, avenues, crossings and alleys to get at one's objectives. In a city, *over there* always seems too far away which calls for determination on the part of the subject who seeks it. Overcoming distances, changing one's way, getting lost and meeting oneself again, straying, searching, departing and finally, arriving seem necessary subjective conditions of urban living so that city corners can be made one's own and become familiar. Therefore, going around in the city has as one of its many purposes broadening young people's subjective possibilities as the spatial dimension is recreated, as well as the subject him/herself and the world. Between comings and goings in the routes of a city, the spaces are being constructed, invested by subjects with elements of a lived past and glimpses of what is yet to come. The spatiality produced becomes, consequently, the extent of history and personal deeds, a result of reminiscences of past achievements and future possibilities.

In this paper I intend to discuss how circulating in the city, representing a condition of social exchange between men and women in a complex urban environment, has its psychic equivalent in the construction of subjectivity by joining temporal dimensions of past, present and future and establishing connections between action and projection. Circulating in the city allows the combination between trajectories and projects, what one is now and what one wants to become. I also discuss how such subjective expansion, favoured by circulating, is subjected to broader controls of urban life: social control via visual consumption régimes, the commodification of goods and the consequent differential appropriation of the wealth of the city. Children and young people of distinct social origins seem to have different choices of trajectories in the city, what creates variability in the process of aggregating values by the choice of such routes. To illustrate the main thrust of the argument, empirical data is provided, obtained through focal discussion groups with children and young people, individual interviews and participant observation of theoretically relevant groups like football team fans.[2]

Circulation and the Construction of Subjective Experience

Circulating in the city is the starting point for getting to know it. Nevertheless, the forms of circulation in the city, the chosen spots where one wants to get to, with whom and when, are yet to be determined by other conditions such as age, imaginary and real belonging to economic and social groups, social status and cultural experience. Therefore, circulating in the city is not done randomly but is conditioned by social interests, values, norms and beliefs sketching the range of circuits that people may take in a city, delineating for each subject a possible city, among many. On the other hand, the city in its multiplicity and diversity of resources, is always alluding to other lives, *other* pleasures and experiences than those consciously consented by the subject, allowing for escape routes and transgressions of initial choices and bringing forth unforeseen possibilities. Consequently, circulation is established as a connecting point between present and future, between what is done and what could have been done, visions of time and space that are superposed and end up linking present and future at the same point in time.

To children and teenagers, dislocation through the city triggers the initiation to the collective processes of social life, the decoding of power relations crystallized in the apparently immobile appearance of space and the complex interface between space and human action. First of all, dislocating means to submit oneself to the concrete solidity of the human body and its limited resources to move around. Dislocating assumes distinct features either by the range of action it may allow or by the available resources at one's disposal. Dislocating in a city like Rio de Janeiro implies *having where to go to* as well as *how to get there.* A great deal of children are limited to the place where they live and do not see any point of moving around because they are not even capable of imagining that the vicinity might present enticing choices to them. Others have no money for bus fares and remain secluded in ghettos - in the communities where they live. Thus, dislocating and circulating through the city comprises an important part of a *psychological capital* (dispositions and interests that make one move around) whereby young people can have access to the *cultural capital,* i.e. the symbolic and the material resources the city has got to offer, such as education, entertainment, culture, among others. Envisaging the need to circulate and being able to put it in practice are subjective dispositions that counterbalance the inhibition and fear to know different places and meet different people. Many times the real impossibility to overcome pecuniary restrictions severely limits the access to other places, so that the subject gets enclosed in spaces that restrict his/her chances of seeing, experiencing and occupying the city. The confinement of young people in ghettos is not always directly related to social and economic disadvantages, since wealthy young people are also restricted to condos and malls.

Dislocation through the city makes visible the enactment of

power and economic relationships. To poor young people the *distance* to the symbolic and material resources of the city (usually concentrated in some areas rather than others) limits their chances to increase their educational and cultural horizons. Consequently, dislocating through the city also means a huge differential when it comes to poor youth, as compared to their better-off counterparts, who have to pay for higher fares or, otherwise, remain where they are. In this sense, for poor youth *more* is demanded from them in terms of psychological resources to face long distances that constitute themselves initial barriers to overcome. For poor children who have not the same freedom as their elders to move around, severe restrictions of access to the cultural and educational equipment cannot be easily compensated. Insulation from ampler social exchanges can have accumulating effects so as to undermine aspirations and identifications with *other* (different) ways of living, attitudes and values.

Dislocation also allows for experiencing the materiality of space, recreating it and giving it meaning through action. In this sense, space would not be felt as inert and inanimate but a condition of human action, part and parcel of what can be acted upon and transformed. Dislocating through the city means acting, the establishment of new beginnings, either by habit, reasonable decision or yet by transgressions (intentional or not), that are multiplied *ad infinitum* and interwoven in the huge net of the city social interactions. Action happens when the subjects, *ones in relation to the others,* create a space of interaction. Edward Soja, a geographer, in the line of a postmodern critical geography, states that:

> "the production of spatiality, along with the construction of history, may be described as a means and an end, the presupposition and the embodiment of action and of the social relationships of its own society. The spatial and social structures are dialectically interlinked in social life, and not only mapped out as categorical projections." (1993: 155)

Soja's studies serve the purpose of warning us about a new

perspective concerning spatiality: space is always created by the action of the subjects among themselves; it cannot be conceived as an abstraction, which just allows for locating things and people. Therefore, space and action are interlinked which helps us to regard dislocation through the city as the means to create interactions among people, either reproducing or changing them.

Children, more than teenagers, dislocate themselves through the city with a lot of restrictions. Their movement is restricted not only by age but also by their inability to deal with the complex mechanisms that the city offers, such as the travel codes of passers-by on the routes and the geographic distribution of bus lines through the city. It is not their age itself that does not permit children an ampler circulation, since a lot of adults, tourists or migrants, also have difficulties although they are not stopped from circulating. To children, circulation is limited because the city, built by adults, reflects the social division of work where the household and the school are the places *where children should be*, whilst the streets, or the city in general, are *not* the places where children should be. Therefore, children are restricted to specific and well-defined places, the household and the school (at most the street adjacent to the house), which reveals their position of minority and their limited access to other relevant social activities in the city. As Jens Qvotrup (1993) states, in addition to school activities, which are certainly constructive, we cannot forget that they are not the only ones, and "whenever children interact and get in touch with nature, society and other people - adults and peers - they are contributing not only to the formation of childhood but also to society" (1993: 11). Even if there is formal acknowledgement to this thesis, the institutionalization of children in such spaces confirms their invisibility in the city, reinforcing the deep-rooted idea that they are not capable of significantly participating in other activities.

The complex spatiality of the city reflects and personifies the will of adults and rarely provides for the desires and interests of the social group children belong to. Detached from participating in building and occupying the spaces of the city, children are limited to the households, the playground, the school, but less and less to the streets. These spaces, those are built *for them* and not *by them*, cannot always be redesigned according to their own desires and interests. Mayumi Sousa Lima, in a seminal work about children and the city, analyzes school spaces in the city of São Paulo, for example, as spaces that materialize the relationships of power between students and teachers, between children and adults. She says:

> "The relationship we hope to establish inside school spaces is not a relationship between equals, meaning that every one affected has a right to express one's opinions, to object, to discover and to learn. Every spatial scheme reflects the relationship of authority and discipline. High windows try to prevent children from curiously gazing at the outside world. The arrangement of the desks, the position of the teacher, all leads to forcing the concentration upon children once uninteresting syllabuses and methods aren't efficient to make them focus." (1989: 58)

Mayumi Lima, an architect, criticizes the pretension of adults, architects, urban planners and politicians of being able to build spaces directed to the desires and interests other than their own, without consulting and listening to what young people have to say.

It is then from the household to school and *vice versa* that children should circumscribe their routes in the streets, without diversion or deviations in as a straight line as possible. Warned by their parents not to wander about and to give up attractions that might drive them away from their paths, they are headed either to their houses or to school. Children learn, mainly from their mothers, that they should not pay attention to strangers, go to *other places*, stay outside for a long time, enjoy themselves with their friends... And the street becomes odd and full of dangers.

Interviews with children showed that their initial range of circulation in the city is limited to their way home from school and *vice versa*. This is their starting point to know the city moving away from home with the parents' agreement; it is a moment full of anxiety both for children and their parents. However, there is also ambivalence on children's part, a mixture of fear and curiosity. In the example below, E., an eight year old boy, has started going to school by himself for a couple of months:

> *I* : How do you feel going to school by yourself?
> *E* : I don't know. Sometimes it's good, I go along with my friends. Sometimes on my own, then it's more boring.
> *I* : Boring, how?
> *E* : My mother keeps telling me to be careful, never to talk to strangers and not to stop anywhere. Then, I keep thinking about it and it's boring.
> *I* : And when you go with your friends?
> *E* : It's cool, because I'm with them. We do other stuff. But I don't tell mom.

Thus, E. feels 'protected' by his friends when he dares to do 'other stuff', whereas, when alone, he seems submitted to his mother's admonitions that hinders deviations and a freer circulation in the city.

By circulating, young people can make the city their place. By the detour of routes commonly known one can get lost in the city and thus obtain from this situation the experience of finding, of (re)discovering what generates new perceptions and knowledge. Wandering about and drifting through the city corresponds to roaming, since having no aim suggests a rejection of purposeful and planned mobility on behalf of the fortuitous and unpredictable. The willingness to get out of one's way and wander already represents a subjective demand for the loosening of family links that attach the individual to the censure of family control and rules (Castro, 1997). In small towns, for instance, wandering is negatively embodied in the stereotyped figure of a hobo on whom it is projected the negative consequences of wandering and of aimless living. To some extent, roaming in the city makes everyone an outsider in his/her own land for one has to leave the familiar and protected environment of home and search for novel experiences. The strangeness of the city portrays, at last, the conflictive nature of personal identity marked by the abyss within every human being who is here and there at the same time, lacerated by multiple times and spaces.

Circulation as Point of Juncture

Trajectories in the city are not random, and this means that the city does not present itself to children and teenagers as an undifferentiated whole (Magnani, 1992). Chosen routes result from choices that allow for certain social and cultural practices (education, entertainment, shopping, among other thing) and the circumscription of the city to the logic of personal history, present and future. Already lived history and that yet to be lived both conjoin to establish the dense moments of the duration of the trajectories, which establish a link between present and future. When young people circulate in the city, they behave as if they are already there, by showing the visible dispositions, attitudes, clothing and badges that convey the psychological proximity with the place where they are going. Here and there are united, as opposing poles divided by the route, pushing time in a paradoxical way: being on the go means not having arrived yet, but also means being already there in an anticipated and vicarious way. Thus, trajectories compress time both by projecting the subject far ahead in a vicarious future, and at the same time delaying the outcome postponing the end to some ulterior moment.

In 'city workshops' with children and youth carried out in different districts of Rio de Janeiro, where they were invited to talk about their routes in the city, the experiences that they had outside home, whom they met and how they found their way as they went to different places, some children of the community of Acari chose "the car" as the element that best portrayed the city of Rio de Janeiro.[3] Interestingly enough, this choice was made by children who had not any experience in using a car, since in this group none of the children had ever ridden a car. Commenting about their choice, children said that *"if one goes shopping, what one has bought can be kept in it"*, or that *"they don't need to stop at any bus stop, going straight home"*, or simply that *"if one rides a car is because one is very smart and clever"*. Their choice is remarkable, not only because they see the car as a useful object, but also that the car permits circulating more easily and freely in the city, and through circulation, accomplishing smart and clever things, consequently being smart and clever oneself. The car stands for the freedom of enjoying the city, of increasing their range of action over space with minimal effort.

In another school,[4] where city workshops were carried out with children of the same age as above, they told of the desire to walk along the streets by themselves. There were very few places where they could go without being escorted by an adult or an older sibling. Only if they wanted to go to a place close to their houses, would they have permission to go by themselves. One of them claimed that she liked to go to the bakery because it was the only place where her parents let her walk by herself. Like the children of Acari, they also experienced a kind of restraint, but their fate was even further privatized: most of their free time was spent inside their own houses, in front of the TV or playing games with the computer. Dislocation, if desired, was not at all approved by parents who, according to children, feared what might happen to them. On the other hand, children excitedly imagined themselves circulating through the city, what they might find out and how they might feel.

One of them addressed the group in this way: *"Imagine coming across someone who has just been run over? I've seen it once through the car window, but it was too far away... Never been close to it."* Another one said: *"But do you think you can face it, to see someone bleeding and run over like that?"* And the other answered: *"Of course..."* Another one who listened said: *"I would like to know... what it is really like."* (he stressed the words *I* and *really* as he talked).

Thus, curiosity to come across different situations and find out *what it is really like* attracts children who are generally severely limited to known and protected spaces. Imagining how it is to circulate freely through the city, children picture themselves and cherish experiences yet to come. The perspective of *here* and of what is known is expanded to accommodate what may be known and everything that is over *there.* Consequently, here and there break with the apparent fixed and disjoint location where they stand to converge towards one same moment in time contaminating it with new and unforeseen experiences, parting with crude facts to accept longing and fantasy.

To young people mainly, circulating in the city is imposed as part of socialization processes which lead them to the insertion in new social groups, especially those of peers. "Young people are going along the streets listening to their walkman. They seem to have forgotten their native land. By the way, what does native land mean?" (Guattari, 1996: 293).

Neither native land, nor original family: young people long for getting lost in the city, so that they can rebuild other references of identification. Meanwhile, such unsettledness which slowly erodes vernacular and family traditions, does not necessarily promote the establishment of new spaces of freedom. Guattari (1996) reminds us that nomadism is captured under the control of capital that introduces, through the production of images, a new pedagogy - a visual one - that would be a form of social control. When young people wander through the city leaving behind them 'familiar' admonitions and expectations to attempt to establish new identifications, they certainly recreate the urban environment in accordance with new values and lifestyles that gradually evolve within urban youth subcultures. However, such changes take place subjected to other dominant points of view, which organizing space through visual codes, socialize young people into specific consuming standards.

Being away from their origins and the obligations dictated by kinship, young people get entertained in the acquaintance of their peer group which now becomes the main and sometimes the sole reference for youngsters. The generalized feeling of not being attached to anything or committed to nothing is celebrated as a new brand of youth's lifestyle. Wandering through the city aimlessly, always eager to deviate in view of unforeseen experiences, avoiding keeping track of time so as not to control the duration or the schedule of activities, these are forms that youth reinvent social bonds. Instead of being located, hovering over every place and making oneself multiple; being everywhere and nowhere at the same time, assuring an unceasing movement that makes it rather difficult even settling down. The chorus of the song below, which has caught on among young people for a long time in many Brazilian cities, deals with the uncertainty of not wanting or not being able to settle down and assume obligations:

> *"I belong to nobody*
> *I belong to everybody*
> *And everybody loves me"*
> ("I know how to flirt", The Tribalists, 2002).

"To belong to everybody" denotes the desire to expand oneself *ad libitum* without anything to inhibit such an expansion. At the same time, "to belong to nobody" denotes the need to give up ties, not to be committed to anything. The apotheosis is love of everybody - the climax of an (impossible) dream of not being subjected to the weight of matter and the incommensurability of space, the motto of a postmodern hero that elects movement and mobility as his/her new proclaimed values.[5]

Moving oneself, displacing oneself from place to place, and circulating increase subjective possibilities at the same time that mobility is itself propelled by the demolishment of local references through globalization processes of the territory. In the aforementioned process Santos (2001; 2002) explains how local values are disrupted by global processes, generating an ever-increasing disorientation and fragmentation. He says:

> "... there are territories where 'an absolute globalization' takes place and others where 'just a relative globalization' happens. In the first case, resulting in a more complete presence of globalization processes, production, circulation, distribution and information tend to be more

> corporation-oriented, that is, depends on enterprises whose vectors are money and its purposes: speculating activities, commodification, exportation activities and so on." (2001: 205)

Santos says that this kind of globalization produces spaces that affect behaviours, as the establishment of such corporations make others (local ones) become disorganized, making space more fragmented and incoherent. Santos warns that circulation today is what withholds the order of values in spatial transformations: enterprises want to offer advantages like entertainment, job positions and wealth so that spaces become attractive. In this way they build images of spaces that are like magnets attracting people. The dialectics of the territory implies the convergence of local and remote controls, the latter away in the global cities that have political control. Santos says: "The result is the acceleration of the process of alienation of spaces and man, one of which components is the huge current mobility of people." (2002: 269)

Circulating becomes then the effect of a re-ordering of local values and references in favour of new paradigms dictated by far-away centres whereby some spaces are made attractive and mobility towards them becomes a must. *"Belonging to everybody"*, as convoked by the aforementioned chorus, means being open to new injunctions of the territory, now extremely unsteady and changeable, that harbours the intents and symbols shaped in far-away places. Therefore, as a result of the rearrangement imposed by capital, space, bonds and sympathy are reconfigured. Young people seem captive in the cultural dynamic that drives them towards the vertiginous mechanism of unsettledness.

If displacement can mean symbolic capital by the access it allows in terms of cultural, educational and entertaining facilities, it has also become a condition of subjectivation in face of the new forms of social control. Young people, more than any other social or age group, are submitted to the visual pedagogy that the city imposes '*making everything resemble*', as Theodor Adorno (1944) once put it. In exerting group nomadism, looking for references other than the family, for spaces of freedom and other possible sociabilities, in a world transformed by technology, young people endlessly consume the 'cultural animation of town', as Arantes (1996) states, that is, ever more syncretic cultural forms, now internationalized and which are due to last but for very little. She says:

> "The inherent ambiguity of the cultural processes of the cities in which the economic relations and relations of power are disguised as cultural differences to be either preserved or cultivated ... is the de-characterization of diversity through the expansion of a market ruled by the principle of serial reproduction, standardized and so forth." (1996: 237)

In this sense, there is absolutely nothing that does not serve the purposes of international capital whereby anything can be commodified, even local values. Thus, the attachment to local references, the feeling that there is a 'home' can eventually cause malaise and be felt as inadequate if the motto is *to belong to everybody*, in this sense divert to whatever catches one's attention and be converted to an internationalized taste.

To be a teenager in a city like Rio de Janeiro implies getting converted to nomadism. When it comes to "taking advantage of life", "being everywhere", "trying every novelty", and above all, "making the best of life", nothing can be left behind. The diversity of the city embellished to serve the taste of the wealthy, mainly in its "sparkling areas",[6] challenges young people, largely those who can circulate and consume new experiences and feelings, producing in them the desire of unceasing movement. Nomadism parallels the unnecessary surplus of attractions, excessive by all means, where young people must be permanently summoned.[7]

To poor young people, however, such nomadism is permeated by constant vigilance by the police apparatus leading to a variety of humiliations of youth and the reinforcement of stereotypes. Cassab (2001), in her research on poor young people circulating in the city, reports that the comings and goings of these young people are dotted with situations in which they are, beforehand, put in a position of offenders and juvenile delinquents, whether they are headed for work or for the ball:

> "In each and every situation, present in everyday life, these young people are subjected to a certain type of surveillance, that identifies them based on certain stereotypes. Such identification does not personalize them, does not retrieve their singularity, on the other hand makes them anonymous." (2001: 43)

Anonymous but recognized in a stereotyped way as dangerous, these young people have to restrict their circulation in the city even if they might be seduced by the countless possibilities and offers that the city has. Having their circulating space severely reduced, their access to certain noble areas of the city becomes an adventure with probable negative consequences. As one of our interviewees said:

> "*M:* I don't hang out at the mall with my friends because some security guards have already told us to leave several times. We "was" not raising a racket. Just having fun. But they look at us and see we don't belong to the same class. (M., 16-year-old teen)."

Consequently, circulating in the city is different for poor young people, generally blacks and mulattos, who, by racial characteristics, are automatically included in a disadvantaged socio-economic group, whenever other visible signs of privilege are also lacking.

Poor young people would be the most immune to the control imposed by the new visual regimes, once they largely circulate in those spaces that are less prone to globalized changes. Being outside the prestigious circuits of town, these young people would be those that Zukin (1993) has described as possibly resisting the dominant values conveyed by visual consumption that lead to conformist identities in a postmodern society. They would have "an anti-socialized regard" more attached to vernacular modes in spite of the market powers. Being unprotected by law and having their capacity of freely circulating in the city diminished, poor young people would represent the "not included" part of a society that claims to be "inclusive" and cosmopolitan.

The speeches of these young people reveal that *"gotta be smart all the time and never be a fool"*, or that they can't *"just keep checking 'chicks' out because people start giving us a weird look"*; the colour of their skin, their dressing code, the way they walk and talk, denote a condition of structural liminarity where they find themselves. Therefore, the restricted nomadism that poor young people seem to exert in the city appeals to more local values. Unlike young people of the middle classes, whose motto in the city is having more and more fun...and making the most of it..., poor young people are defied to study and to work, as a way to be included in the consuming circuits. Or else, the alternative presented to them is to assume a deviant course of action and an identity of outlaws. The former imply routes in the city that reveal projects of inclusion as a way of restoring associations and local values. One of interviewees said:

> "I leave home early everyday, at 5 a.m. I take a train and go to work. I work at a paper plant. When I leave work at 5 p.m., I catch a bus and go to an evening school, high school. I stay there until 10 p.m. Get home, have dinner and go to bed, at 11, 11:30 p.m. Everyday is the same. Can't do anything else. But I dream of a better life some day..." (S., 18-year-old teen).

In general, young people do not want to move to another place, but improve their living conditions, in sanitation, transportation and entertainment. They wish they can "get to the top" whose acknowledgement basically depends on their personal efforts to overcome the disadvantages of their position. In the stories that they have told us, they envisage education as the solution to such difficulties as reported in the stories made up along discussion-groups with them:

> "A young man couldn't count on anyone's help to fulfil his dream, not even his family, since his parents had abandoned him and had an alcoholic addiction; the solution to his problem would be enrolling in a school."

Or yet :

> "Lucas, an 18-year-old young man, lives with his family; his mother passed away when he was 15 years old, then he helps to support the family. During the day, he's a labour worker in a building site. At night he goes to high school; in weekends he takes professional courses and is very lonely. He doesn't have time for friends; he's always studying and doesn't give up his dream of becoming a great psychologist; His willpower will help him make his dream come true and will help others, who like him, share a sad history."

Thus, it is on their way to school, to work and back home that poor young people picture themselves in an eventual position of less suffering, bringing hope to their present. Values like having a family, a job that provides them with a worthy life, having a home are reassured, nourishing the fantasy that the city has "good" and "bad" paths and that each person has to choose whichever will lead to what one wants to get at. By choosing 'good' paths, both in a literal and a symbolic way, young people feel they are nearer their high-valued objectives thus enhancing their social esteem in face of an adverse present.

Between Cila and Caribde, between alliances and local values and the nomadic existence that socializes and gives them a globalized taste, young people and children keep dislocating in the city, redesigning it through the routes they can take. Routes that refer imaginarily to projects and the future, creating another *city* within the city itself, built by the constant digression of the present time. The school, the workplace or the entertainment, all contribute that young people and children establish *other possible paths*, ones that enlarge and expand the present time.

The city, as portrayed as prodigal in pleasures or as a pole of contemporary, social and material exchange, is frequently considered a myth, shown as "a good object", abundant and whole, capable of providing full satisfaction. In city workshops held in Rio de Janeiro, young people and children pictured the city as *"good chow", "good stuff to eat like lollypops and hotdogs"*, and, sometimes, *"stuff you can easily get"*. The city becomes the gift to be cherished without delay. Florido (2001), reporting about the city workshops that she conducted in another Brazilian city, S. José dos Campos in the state of São Paulo, also got statements from young people who imagined the city as *"the generous teats"* of a woman. Thus, it is in the city, mainly in the streets, where satisfaction is searched for setting in motion the imaginary of pleasure at all costs. By *"diverse"*, etymologically *"diversu"* in Latin, which means change of course, deviation, one craves for finding what is missing in the present.

Diversion would be the search for the delightful deviation in face of the sadness of human life. Reading the literary work of the French philosopher of the eighteenth century, Blaise Pascal, Bittencourt (2002) analyzes the meaning of "diversion" nowadays. She launches the hypothesis that diversion becomes the possibility to watch and enjoy the pleasures provided by the consumer culture as one succumbs to the dismay and distress of death and human fragility. The author quotes Pascal, "been unable to cure death, poverty and ignorance, men have decided not to think about them, in order to be happy. Diversion is the only comfort for our misery" (quoted in Bittencourt, 2002: 104). Diversion serves the purpose of solving human dilemma facing its own end and suffering.

Diversion is considered by young people as going beyond the boundaries. First of all, the boundaries of protected space, like going out and away from the family and the familiar. Secondly, social ones, like getting to know other places and people, connecting to other identifiable references. In both cases, the issue at stake is the fact that mobility becomes the self-foundational in itself. Diverting and displacing imbricate establishing a common ground that shapes adolescent subjectivity nowadays. Ian Watt (1996), who wrote the stimulating study on modern individualism called *"Myths of Modern Individualism"*, shows the characters of Don Quixote, Faust and Don Juan, who personify basic characteristics of modern individualism, as *"dedicated travellers...lonely nomads"*. The three characters revert the trend established since ancient times that "going out" would be equivalent to an exile, a curse. On the other hand, such characters seemed untied to either family or domestic attachments, whose lives were based on unceasing mobility and on travelling as a means of experiencing new pleasures, new challenges and knowledge. Quoting Watt:

> "Against this perspective, Faust, as we know, travelled around the world, as far as hell. In the adventures of Don Quixote his first act was to leave home and let luck and Rocinante determine his way; whereas Don Juan travels to Naples and back and then to several places in Spain." (1996 : 122)

In this way, displacing and circulating are associated with solitude, since they are centred on the movement of "the self", who invests in the detachment and distinction in relation to others. As their literary counterparts, young people, modern individualists of present time, act to accomplish their individual callings, trying to be themselves to the detriment (a lot of times) of the collective and family appeals.

Circulating serves as a metaphor to the practice of diversion, enhancing the movement and the unceasing search, emphasizing the individualist and lonely sense of transgression. Diverting oneself might be considered the reconciliation of the subject with society, the conciliation of tensions and elimination of conflicts. Already in the 50s, in the last century, Adorno admonished, "diverting oneself is being in accordance with". He goes on:

> "Diverting always means not having to think about it, forgetting, suffering, even where it is shown. Impotence is its base. It's in fact a escape, but not as stated, a escape from a bad reality, but from the final idea of resistance that such reality still provokes. The freedom promised by diversion is the freedom of thought as denial." (1944: 135)

Through diversion one gets in touch with an endless variety of assets portrayed by the capital order as enjoyable. Diverting cannot happen randomly but under the spotlight of the consumer culture that dictates where one stands out through dominant styles of life and delight. Diversion opposes distancing from and resisting to the social order, outlining a necessary communion between the subject and society, as a kind of oblivion is required to obtain relaxation. On the other hand, the variety that unceasingly challenges young people and children inviting them to delight demands experimentation free from moral obstacles that can put limits to what can be done. Thus, circulating and diverting become intrinsic conditions of an individualist freedom based on the principle of "I want"= "I can" (Berlin, 1969).

M., 14-year-old girl, a resident of Lins, says:

> "I do whatever I can to have fun… Stuff my parents don't imagine or even suspect…"

L., 15-year-old girl, a resident of Meier, says:

> "I go to a mall called "The Third Millennium" when I hang out with my friends. When I can, I go to a party at "Urca Hill", then I get together with any guy pretending he's my boyfriend to be able to get in…"

A., 15-year-old girl, a resident of Cordovil, says:

> "We got to go out to see stuff… making the most to go to the best places…"

F., 17-year-old girl, a resident of Penha, says:

> "Since I was little I like to hang out and get to know places. Nobody forbids me... I do it cause I like it, there's nothing I can't do. I give everything a try. It's good. It's freedom ..." [8]

Circulating allows facing dangers, calculable to a certain extent, that might incite and divert young people, leading them to enjoy *intensities*. Intense emotions do not belong only to ritualistic and collective moments as parties and carnival anymore. They have become part of everyday life, when, at any moment, one wants to experience the state of excitement connected to the dimension of individual freedom where sensuality and pleasure play a significant role. Daily life and partying meet, making relatively undifferentiated the order of the profane and that of the sacred. The search for intensities aims at overcoming boundaries towards an abnormal - hyper state - so that the normal state of affairs becomes shaken. In this way diverting nowadays assumes a transgressing connotation. Living with "a rush of adrenaline" has become the *leitmotiv*, which should increase the intensity of the half-hearted existence of daily life. Extreme sports like bungee jumping, skydiving, rappelling, cross country racing, etc. that use movement and speed as elements, are pursued in order to provide the excitement that breaks daily routine. Falk (1994) emphasizes how corporeality and transgression interrelate in the contemporary time, when the limits of the self might dissolve in the pleasurable experiences of bodily functions. The distinctions between daily life and partying, profane and sacred, are faded since everyday can be a "party day", and it seems merely an individual decision, and not a collective and ritualized one, of how and when each one can have fun. Above all, in the current scenario, diversion might be consumed *ad infinitum*, without any restrictions, in fact, nourished like *an industry*, as Adorno would say.

Thus, circulating, diverting and transgressing are interwoven in the post-modern scenario mutually determining one another. Our ethnographic data on organized football team fans[9] contains various passages where circulating, diverting and transgressing become interchangeable aspects in the search for intensities propitiated by risk experiences in the streets of the cities. Here, a small passage from field notes is transcribed :

> "I got to Largo do Machado about ten past twelve, on time. There were about 15/20 people. I sat on the edge of the public fountain and started talking to one of them about the other game, the week before. 'Flamengo' had lost that match. He told me about his hopes for today's game. 'Flamengo' has to win 'Vasco', he said, by two goals to become the champion. Suddenly I heard a noise (splash). I looked back and noticed that stones were falling in the fountain. I raised my eyes and saw a group of people coming out of a street moving towards us. Then, one of the guys who was with us said: "Nobody runs, let them come by, they're not strangers, let's see what's about!". Meanwhile, another big stone hit the foot of a person next to me. I looked ahead, some guys had already turned the corner about 170 feet away. I also started to run, to the blast of a rocket, a kind of firework (which was lit towards us) and, as I heard of later, shots (two) fired into the air. I got into a supermarket and stayed there for a while with the other guys. At that moment, people's reaction surprised me. They were sympathetic and even helpful, willing to give us assistance if they could. Criticizing the attitude of the "vascaínos", the fans of "Vasco", but they didn't seem to realize that we were also part of the mob. The fans of "Flamengo" are also given commands to strategically drive the fans of "Vasco" into a corner. The situation wasn't the reverse due to the occasion.
>
> Going back to the public square each guy had a different version of the episode. One of them, who decided not to flee, faced the opposing group and showed us his injuries.

He was a rather tough guy, who cried while showing the lamps on his neck and back, which were pretty scratched. His lips were very swollen and his legs very bitten. Even limping a lot he didn't abandon the group. He didn't seem to be over 20 years old. At that moment, one of the guys who was returning, said he had seen three rivals talking nearby. They made gestures showing how they had beaten up one of the rivals. They decided to chase him and I, in spite of the fear, went along. The passersby had two distinct kinds of reaction: they either criticized them "Gosh, all of this because of a game", or incited them, like and elderly with a cane, "So, are you going to beat up the 'vascaínos'. Wait up 'cause I will help you...' ". They got into a restaurant and were protected by a security guard that stood at the door. The hero of this episode was a guy who broke into a restaurant, beat up a "vascaíno" and took his shirt away. At that moment, I was on the sidewalk across the street, like the other guys. Just ten guys were willing to go into the restaurant. When the guy (who wasn't even tough, and was apparently only 16 years old) left the restaurant showing everybody the shirt, everyone cheered him and clapped hands.

We were heading for the square again. The place we had been was too close and soon we reached it. One of the leaders talked about our group needing a gun. The other guy said they should have all broken into the restaurant to kill the "vascaínos". "We had to break everything. Later put it on our tag". The guy who was beaten up claimed for the ownership of the stolen shirt. He said he wouldn't sell it. He would keep it as a token of their victory. He said he had recognized of one his aggressors, "We used to study together, I don't care, I wanna kill that son of a bitch!". They were enraged and seemed to be immune to the action of the police. They do not care about the consequences."

In this fragment, the street has become a place for celebration, a mixture of flight and attacking behaviours, in-group solidarity and euphoria. Physical and moral attack of the supposed enemies provides enough adrenaline for intense emotions. The feelings of rage towards the enemy, in this case, the fans of the opposing football team, furnish the necessary motivation to the emotional breakout through quarrelling, cursing and fighting. Corporeality is experimented through intense feelings of fear, anger and pleasure built up along the unceasing movements - wandering, running, sweating, grasping, fighting, hollering, escaping, hiding and chasing. Time becomes saturated and engrossed, alien to tranquillity and rest. Such corporeality supports sociabilities among young people woven through such fortuitous encounters guaranteed by the display of badges and common symbols such as clothing, the way one walks and command words. Circulating leads to spending energy, using up time "doing stuff", being busy anyway opposing any situation that alludes to quietness, once the former seems to be experienced as void and boredom.

Therefore, to be quiet and still, potentially arouses boredom, as for example, when the subject is alone on a means of transportation in the city. On a bus, train or any means of transportation, one spends a long time idling, being submitted to a forced immobility and a mute proximity with strangers. This inertia can mobilize the feeling of void by "not doing anything". Children and teenagers describe such situation as *boring* and even, unbearable, because they feel captive, locked, stuck to a place that can't leave. When asked about their thoughts during a bus trip, they said *"they don't think of anything"*, or that *"they think about a place they wish they knew"*, *"in pollution"*, or yet *"what worries them, like homework"*. The trip becomes a long and distressing time in their lives, as the "naked man" appears stripped away of his/her rush and hectic routine. So, for this reason, in order to avoid the void, the boredom and the 'nakedness', bus trips frequently become the perfect occasion for all kinds of agitation whenever children and young people are together: they tease other passengers, shout, sing, tease passersby at bus stops, curse, joke, mess everything up, in short, they manage to have a lot of fun during this time; for them meaningless duration, time-void, time-abyss, time-immateriality, time-death. Without having anything to do, one runs the risk of being nothing, or not being able to do anything, not spending, not consuming that is being reduced to nothing.

Diversion that Lasts : 'Conversion' is Possible

Circulating in the city portrays to each subject a real and a possible city. If the real city is constituted by the routes effectively taken, they are always destabilized by desire and identifications with what is not present but can be imagined; with what belongs to others, but can become ours. Circulating in the city does not happen randomly but according with an internal compass that allows one to aggregate values. On the other hand, the city does not belong to us entirely. It holds the cosmos that always offers itself partially, sides of light and sides of shadow. However, the spatiality of each city is extremely variable since to some people the real and the possible city are portrayed in a timid way. To lots of children and youth, living in ghetto areas, it is not possible to get to know other places. For example, in the city workshops held in Acari all children complained that they had never left Acari, they had never got to know any of the other suburbs of Rio de Janeiro. They had never gone to the beach, knowing it only through the TV. They had never ridden a car. They had never gone to any mall or to the movies. To others, the city had become a taboo, due to the fear of facing different places. To M., a 16-year-old boy, living in Meier, the city can only be visited by car, not only because is it more comfortable, but also *"because you don't run the risk of getting lost and ending up in a slum"*. To this youngster, *"Rio does not offer good places, me, I'd like to know the world, just the world..."* Odd identification with the world, since the fear of different people freaks him out.

Wandering through the city outlines the different ways that young people find to build up their own history. This is certainly conditioned by their social and economic backgrounds, but also depends on their subjective dispositions to face the new and the unpredictable. To a lot of young people, the routes become held back, blocked and left out by the fear of violence and by avoiding the places they regard as dangerous; to others, the routes show conversions or changes from one way of being to another. A., a 16-year-old girl who lives in Catete, reports what she felt like when she caught herself circulating in the city more freely, able to choose her own paths:

> "I started doing other stuff. Suddenly, my life changed, I stopped watching too much TV as I used to before. Started taking acting lessons, hanging out at the school science club. I go everywhere and know more people. Been thinking more about the future, my life, a profession."

The "suddenly" A. refers to, alludes to what is imminent, but that must be activated and welcomed by the subject. The possible city is that which can be updated by the action of the subject.

The possible city which blossoms side by side with the real one, is also present in the speech of E., an 8-year-old boy, who lives in Vila Isabel. His clever speech surprises me, not only for his élan, but also in contrast to the depressing and sinister place where he lives: open sewage, shacks without a bathroom and without piped water, where the families live all together, without anything to soothe the deterioration of the place. E. likes the place where he lives except for the shooting and the gutter that turns everything *muddy*. He would rather stay where he lives then going to a better place where his mother wants to go: *"this place is full of cousins and people I know"*. He has been to the mall, ridden in a car and liked it a lot. He says he knows São Paulo and the U.S., and soon explains: because he has seen it on TV. He likes to talk about places, far-away places. His father's statements make the far-away places seem closer and more interesting. He has also been to Paquetá and *"crossed underneath the Rio-Niterói bridge"* with his mother and siblings. His mother was scared to do it, but he wasn't. This experience meant opening up new horizons, and he was not afraid. He says he wants to go to Aparecida do Norte, where there is an amusement park. He likes to go to his friend's house in Bangu.

Also to "Campo de Santana", in spite of not going there as much as he wants.

E. stood out as one of our young interviewees full of life and stories about his real and virtual trips - the ones he took and the ones he dreams of taking. All of them meant not only movement and circulation to other corners, but also mainly circulation of affect and emotion. Moving along to all of these places has provided new experiences that might be told to others: to me, who interviewed him, his father to whom he talks about what he sees on TV; his cousins who play with him. E. appears as a little traveller who has been converted not to the dazzling movement stimulated by consumer novelties (even because he cannot afford them), but to the possible inner movement that makes present time float, making it indecisive and open, updating what he did to the flavour of what is being experienced and of what is yet to come. This way, E. manages to change his tough daily life of fetid muddy surroundings or a house without any comfort to a less timid present where the evocations of previous routes are constantly circulating. If the gutter is fetid and gets in the way of the fun, on the other hand, what would life be like without his cousins? He might have ridden a car just once or just passed by the mall. But both experiences are inexhaustible sources once he faces the past not as something that passed, but as the means whereby the present can be permanently derivated, constantly rebuilt and resignified.

Milton Santos (2001: 261) says that there is a need to "implement conditions to improve the circulation of men, products, goods, money, information, orders, etc." The territory comprises distinct zones of "density" or of "rarefaction" as for the presence of things, goods, people, information and assets, portraying denser or lesser availabilities of information and other social and cultural resources. Little flow would mean restriction as to the circulation towards the access to cultural assets. However, Santos also warns about an "unnecessary circulation" that would bring excessiveness, and through it, an enormous social cost; that is, circulation should not be held as a means in itself, but becomes important only if it allows social exchange that assures the welfare of everybody (in a city, in the world).

Circulation does not always allow the exchange between people. It can be determined by individual displacing, often seeking for news and fun, which works for the consuming of new emotions. What social cost would that kind of circulation bring? Such a question should be asked in face of the proliferation of exchanges that take place but do no provide the conditions for converting subjects into other, different points of view for more than one or two minutes. Or even, exchanges that are restricted to some subjects that benefit from them to the disadvantage of others.

The importance of circulation to the psychological structure of young people and children can be ascertained if we understand circulation as an instrument of expanding subjective dispositions. Most probably, circulation as nomadism and unsettledness modelled by the ever increasing commodification and diversity of consumer culture, cannot entirely consolidate inner changes and psychological transformation. Circulation might be conceived otherwise as conversion: to other places, references and point of views having in mind the possibilities that the city can operate. Thereby lasting significance is attached to novel experiences when the subject can have the necessary time to elaborate, feel and exchange emotional contact so that 'outer' circulation can also mean 'inner' circulation and circulation 'with others'. Otherwise, circulation runs the risk of dispersing itself as an aimless expenditure without subjective transformation.

NOTES

1. Modified version of the chapter *Circulating in the city of Rio de Janeiro,* to appear in the forthcoming book *The Urban Adventure: Children and Youth in Rio de Janeiro,* by the same author.
2. The author would like to thank the support of Fundação Carlos Chagas Fo de Amparo à Pesquisa do Estado do Rio de Janeiro (FAPERJ) and the Conselho Nacional de Desenvolvimento Científico e Tecnológico (CNPQ).
3. In this workgroup children were asked to choose among different newspaper and magazine clippings, spread out before them, one that they mostly identified as something that characterised the city where they lived. The idea was to start discussing about the way they envisaged the city where they live, what meanings came along with their experience in the city, what feelings were

derived from their speech. City workshops were carried out in 8 different schools, in different areas of Rio de Janeiro, better off and less well-off, with children and youth of different ages. They comprised discussions lasting for about an hour and a half weekly, during three months.

4. Pedro II school, a state school in a southern district of the city of Rio de Janeiro.
5. Examples of this 'existential' profile are given by extreme sports like bungee jumping, skydiving, rappeling, etc., or yet experiencing attractions at amusement parks, as the roller coaster and others, seeking, among other emotions, vertiginous movement that causes a new perception of body and space by changing its weight and stability in propulsion and speed.
6. Again Milton Santos is our source of inspiration here. He mentions 'bright spaces' against 'opaque spaces'. The former accumulate technical and informational densities, attracting capital, technology, organization, See M. Santos. *op. cit.*, 2001.
7. Here I agree with Terry Eagleton when he stays: *"Postmodern culture, conversely, is classless, meaning that consumerism is classless, which is equivalent to say that it cuts across class divisions while driving a system of production which finds such divisions as indispensable. In any case, the consumption of classless culture is nowadays increasingly the mark of middle class."* (In *The Idea of Culture*, London: Sage, 2000, pp. 125).
8. Interviews were carried out from March 1999 to December 2002.
9. Final report about football team organized fan clubs by Bruno Pizzi, who from March to November 2001 observed and participated in these fan clubs.

REFERENCES

Adorno, T. 1944 [1986]. *A Dialética do Esclarecimento.* Rio de Janeiro: Zahar.

Arantes, O. 1996. "Cultura da Cidade: animação sem frase". *Revista do Patrimônio,* 24: 229-240.

Berlin, I. 1969. "Two Concepts of Liberty", (pp.118-172) in I. Berlin, *Four Essays on Liberty*, New York: Oxford University Press.

Bittencourt, M.I.F. 2002. *Ilusão e Criação na Sociedade de Consumo: um estudo sobre o divertimento.* Doctoral thesis, PUC-Rio, November, 2002.

Cassab, M.A.T. 2001. *Jovens Pobres e o Futuro.* Niterói: Intertexto.

Castro, L.R. 1997. "O que querem as adolescentes da cidade", *Arquivos Brasileiros de Psicologia,* 49(4): 79-93.

Castro, L.R. 2001. "Crianças, jovens e cidades: vicissitudes da convivência, destinos da cidadania", (pp.113-156) in L.R. Castro (ed.) *Subjetividade e Cidadania: um estudo com crianças e jovens em três cidades brasileiras.* Rio de Janeiro: 7 Letras/Faperj.

Eagleton, T. 2000. *The Idea of Culture.* London: Routledge.

Falk, P. 1994. *The Consuming Body*. London: Sage.

Florido, F. 2001. "Cidade das tetas fartas, cidade das vacas magras: encontros com adolescentes em S. José dos Campos", (pp. 63-112) in L.R. Castro (ed.), *Subjectividade e Cidadania: Um estudo com crianças e jovens em três cidades brasileiras*. Rio de Janeiro: 7 Letras/Faperj.

Guattari, F. 1996. "Restauração da Paisagem Urbana", *Revista do Patrimônio*, 24: 293-300.

Lima, M.S. 1989. *A Cidade e a Criança*. São Paulo: Nobel.

Qvotrup, J. 1993. Nine Theses about "Childhood as a Social Phenomenon", (pp. 11-19) in Childhood *as a Social Phenomenon: Lessons from an International Project*. European Centre for Social Welfare Policy and Research: Eurosocial Report, 47.

Santos, M. 2002. *A Natureza do Espaço*. São Paulo: Edusp, 4ª ed.

Santos, M. 2001. *O Brasil. Território e Sociedade no início do século XXI*. Rio de Janeiro: Record.

Soja, E. 1993. *Geografias Pós-modernas*. Rio de Janeiro: Zahar.

Watt, I. 1996. *Myths of Modern Individualism*. Cambridge: Cambridge University Press.

Zukin, S. 1993. "Postmodern urban landscapes: mapping culture and power", (pp.221-246) in S. Lasch & J. Friedman (eds.), *Modernity and Identity*. Oxford: Blackwell.

Chapter 3

No Ball Games

Children's Experiences of Urban Space in an English Town

Virginia Morrow

The last half of the 20th century has seen increasing urbanisation and suburbanisation in the UK, particularly in the South East of England. Housing was provided on a very large scale, and 'New Towns', suburban sprawl, roads and motorways, have developed rapidly. Children often constitute a high proportion of the population of the suburbs and 'New Towns'. Local authority planning provision does not appear to have been able to keep up with changes in the structure of the population, nor changes in the physical structure of the landscape in terms of provision of leisure spaces, parks, and places for children to play in or to 'hang out' in (Morrow, 2001a). However, renewed focus on the importance of neighbourhoods and communities in UK social policy has generated research that has explored people's concerns about their localities, and a consistent theme that emerges from this research (and policy concern) is anxiety about children and young people in neighbourhoods (SEU, 2000).

This paper draws on data collected in a research project conducted for the Health Education Authority that explored the relevance of Putnam's (1993) concept of 'social capital' in relation to children. 'Social capital' consists of the following features: social and community networks, civic engagement or participation; community identity and sense of belonging; and norms of co-operation, reciprocity and trust of others within the community (Putnam, 1993). The premise is that levels of 'social capital' in a community have an important effect on people's well-being. Health behaviours and practices may superficially appear to be a private matter for the individual, but in reality health practices take place in a range of social arenas, which for children, are constrained by everyday contexts, which will vary from school or institution (for previous research see Mayall, 1994), family (Backett, 1992; Brannen *et al.*, 1994), and peer group (see for example Michell, 1997; Pavis *et al.*, 1996, 1997). Neighbourhood health effects are less well-documented in the UK, where health promotion research with children and young people has tended to focus on individual topics or risk behaviours (such as smoking, drug-taking, and alcohol consumption), and little is known about children's social networks, their use of neighbourhood spaces, their levels of trust and community identity, and the implications of these for quality of life or well-being. 'Social capital' is a concept that has been contested at a number of levels, conceptually, methodologically, and theoretically (Morrow, 1999, 2001a & c). This paper focuses on urban spaces, and how such spaces are used by, and perceived by children.

Methodology and Methods

My research drew on the theoretical paradigm proposed by the British social anthropologists James and Prout (1997). They argue that we need to move beyond psychologically-based models of childhood as a period of socialisation, and emphasise that children are active social agents who shape the structures and processes around them (at least at the micro-level) and whose social relationships are worthy of study in their own right. It is now acknowledged that "research about children's lives is... essential if policies and programmes are to become more responsive and relevant to their concerns and needs" (Boyden and Ennew, 1997:10). A research project was designed to explore the following questions with children:

- *social networks:* what is the composition, durability, ease of access to and frequency of use of children's social networks? How are these networks defined and what do these networks provide, and how does this differ according to age and gender? What does friendship 'mean' to this age group?

- *local identity:* do children have a sense of belonging and identity with their neighbourhoods/communities?
- *attitudes to institutions and facilities in the community:* what physical spaces, such as parks, streets, leisure centres, clubs used for social interaction, are available to and used by children? Do children feel safe in their neighbourhoods?
- *community and civic engagement:* to what extent do children engage in local community activities? To what extent do they feel they participate in community and institutional decision-making?

The research was based on 12-15 year olds' subjective accounts of their neighbourhoods, their quality of life, the nature of their social networks, and their participation in their communities (Morrow, 2000a, 2001a)[1]. The research was carried out in two schools in relatively deprived wards in 'Springtown' (all names in this paper are pseudonyms). Springtown is about 30 miles from London, and has grown very rapidly. The population of the town was 172,000 in the 1991 Census, and the 1996 mid-year estimate was 182,000. Almost one-third of the population of the town are under 20 years of age. Research was carried out in two parts of the town, 'West' and 'Hill'. 'West' consisted of 'suburban sprawl' on the outskirts of the town, with post-war housing and factories; 'Hill' consisted of a mixture of industrial development, and Victorian, inter-war and post-war housing development. Research was carried out in schools, and the sample included 101 boys and girls in two age bands: 12-13 year olds and 14-15 year olds, with a significant proportion from minority ethnic groups. In West School, three children were African Caribbean, two South Asian, three white/ African-Caribbean. In Hill School, there were children whose families originated in Greece, Turkey, Cyprus, South Asia, African-

Caribbean. One or two children declined to describe themselves as being from a particular minority ethnic group.

The research used a variety of qualitative methods:

- children's written accounts of out-of-school activities, their descriptions of people who were important to them and why; their brainstormed responses to the questions 'what is a friend?' and 'what are friends for?', their future aspirations and whether or not they knew someone who was already doing that kind of thing; their descriptions of where they felt they 'belonged';
- visual methods, including map drawing and photography by the participants of 'places that are important to me' (this generated 17 maps or drawings and 100 photos);
- group discussions, exploring use of and perceptions of neighbourhoods, how they would improve their neighbourhoods, and their community and institutional participation. As a 'prompt' in group discussions, children were shown two newspaper cuttings: one depicting their town in negative terms, and one on 'child curfews'[2].

The research thus generated a range of qualitative data, and this paper draws on some of these data to explore children's accounts of urban space.

Friendship and Place: Informal Social Networks

Friends' houses featured in the maps and photos. As noted, children were asked to photograph places, not people, but as the following comments illustrate they often found ways of incorporating friendship into these photos and maps:

> "this road is the most important road because it leads to my *friend*'s house" (Tom)
> "My old school: I often play football there with my *friends*" (Bob)

> "This is our school playground, we hang around with our *friends* there" (Wendy, Leila, Chloe, West area)
> "This is McDonalds. I always go there every week with my *friends*" (Jennifer, West area)
> "Park. Where I used to go as a child and still go with my *friends* now" (Maggie, Hill area)
> "School. Where we sit at lunch time" (Maria, Hill area)
> "I took this photo of my *friend*'s house, because she is my next door neighbour, and I hang around in and outside her house" (Gemma, Hill area)
> "Corner of a street where I meet my *friends*" (Mary, Hill area)
> "[High Road] area. Lots of my *friends* live there." "My *friend*'s house, I'm often in there, and there is a park behind the house" (Jagu, Hill area)

Children often spend more time with their friends than they do with their families, especially as they get older. As Veronica put it: "Why are my friends important? - because I spend nearly all of my time with them". And Maria wrote "My friends are also very important to me because I spend so much of my life with them". Daily activities were often structured around encounters with friends: Maria described how:

> "after school: I often ring my friends and talk on the phone for quite a while as well... After eating I will either go round a friend's house, go to a local youth club, the cinema, or just stay in. My weekends are usually spent in the town centre with different friends. I sometimes go to the cinema, or bowling, or just out."

In many cases, how children felt about where they live seemed to depend on physical proximity to friends: as Maggie, age 15, Hill Area, put it: "I love my house and my area, because there are three parks near me, the town is a five minute walk away, the school is close and I can visit my friends without having to take a bus or walk miles. Most of my friends live in Hill Ward, or my area". Not having friends living nearby was a reason children gave for disliking their neighbourhoods, and this seemed to be more marked in West School which, as noted above, was in a sprawling, suburban locality with few facilities for children. It seems likely that density of population has implications for children's social networks in West School. It was also mostly girls who described this. For example, Olanda, age 14, described how "I'm fairly happy with where I live but would rather live in my old house ... this is because a lot of my close friends live up there. Usually I walk up there most days after school. It would be a lot less hassle if I lived up there near them". Rebecca, age 13, described how she doesn't like her neighbourhood:

> "It's boring, there's not many people of my age living round there. Because my best friend moved away she only lives 10 minutes away, but it's too much to walk everyday there. I've been best friend with her all my life, and I've never broken up with her once. We do a lot of things together..."

Jody, age 14, described how: "It's a bit awkward because I live 2 miles away from the school and my mates live quite far away. At least I still have my old mates who live near me and I always make time for them." In Hill School, some children did comment that they don't see their friends as often as they like, for example, Tobi wrote "During the week I don't go out much because all my friends live up in [Hill area] and I live down town"; and Amanda, who lives outside Hill area, described how, "There is nowhere around the area where I live to go in the evenings, and it is too far to go to my friend's house every night. So by the time I get there, I have to go soon after." Paris, who lived in Hill area, said (in discussion when asked 'how would you improve your area?') "... have all my friends live next door to me".

Ethnicity was also a factor in how children experienced their area. In a group discussion with younger boys in Hill area, two boys said how much they liked their neighbourhoods. Wassef, (of Pakistani-Kashmiri origin) said:

> "yeah, but the thing is, he lives in a good place, in Hill Ward, because he's got all his mates there, it's got a community, and he [the other boy] lives just off town, he's got all the mates you can have, there, he's like the centre of attention where he lives".

Ajit (of Indian origin) remarked:

> "I don't like the area I live in, as many of my friends live far away from my area and I can't walk there, I would rather live in [another part of town] as more of my Indian friends live there."

Children's accounts of where they feel they 'belong' also highlighted the importance of social relationships, rather than geographical place: most said they felt they 'belonged' at home, with their friends and family, and as one boy put it, "I think I belong in a community where I am treated right, and a place that is warm and friendly" (Rock, Jamaican/white origin).

How did the material circumstances of their physical environments impact on their relationships with their friends? The rest of this paper explores children's perspectives on these environmental settings, and highlights the contradictory nature of their experiences of communal areas. While their friendships were mostly experienced very positively, their accounts of practical problems they encounter in leading their social lives in their town and their neighbourhoods, the lack of appropriate places for them to go, especially as they become older, were generally negative. Secondly, it explores their suggestions for how their neighbourhoods might be improved; and, thirdly, it discusses their lack of participation in their communities. As many writers have pointed out, children need chances to do things away from the gaze and control of adults, and the lack of these opportunities is likely to have adverse effects upon their social and emotional development as well as their quality-of-life in the here and now (Adams, 1995; see also Henderson, 1995; Davis and Jones, 1996, 1997).

Children's Perspectives on their Neighbourhoods and Public Space

On the whole, children perceived their town and neighbourhood environments as unsatisfactory. This did vary, in that if they lived in a cul-de-sac in a suburban area, it was 'too quiet and boring', if they lived on a busy street 'there's too much traffic'. There were differences according to gender, ethnicity, and age. None of the children described themselves as a 'Springtonian'. They were well aware of the negative reputation of their town and of particular parts of the town, and which parts were safe and which weren't. Overall, they described a strong sense of exclusion on a number of levels, because of a lack of appropriate facilities and places to go; cost barriers; and practical problems caused by traffic.

Facilities and Activities

Firstly, all children described not having enough to do in terms of appropriate facilities, activities and places to go. Very few described involvement in organised voluntary activities - six boys mentioned being members of a formal sports team; and ten children mentioned using the local youth clubs. For the most children, especially boys, their activities outside school centred around playing football, cricket and cycling, informally with friends in local parks and neighbourhood spaces. However, these parks were often not pleasant places to go for a number of reasons. Harry described how:

> "there's a park where we live, we call it 'Motorway Field' because it's right by the motorway, and it's just covered in dogs' muck, you just don't like to go there, people let their dogs go anywhere, so we like to play football there, but 'cos you don't know where the dogs muck is, you don't play because you don't want to get covered in it."

Dog mess was nearly always mentioned as a problem in the

context of playing in local parks. Rock photographed a primary school playing field near where he lived, and described how he plays football there: "It's good, because there are no dogs, and no adults screaming at you to stop".

Several younger children described a lack of wild places to 'make dens'. One girl in West area mentioned 'we used to have a den, in the woods there, and me and my friend found loads of, like, drugs and stuff, packets and things, so we took them to the police'. One 12 year old boy mentioned that he didn't like his area:

> "cos it's so built up, there's not much to do, and like, where my sister lives, she lives in [another town], and just across the road there's a big forest, and my brother likes to go over there with their dog, and they'd be out for hours and hours, and that's what I like when I go there."

Local youth clubs were seen by older children as suitable for younger children, though a few of the younger children did use them and liked them. Youth clubs were not particularly popular with girls, and some felt there was not enough for girls to do at the respective youth clubs: "all they do is play football and basketball". On the other hand some of the boys in the same year group did use the youth club. Fred described how: "It's open three times a week, I go there most times, sports, play games, listen to music". A lack of things to do and places to go was a general explanation by participants for 'anti-social' behaviour: "we're only getting into trouble because we're hanging around on the street because there's nothing for us to do."

Many children mentioned an explicit exclusion they felt that was encapsulated by 'No Ball Games' signs. These prevented them from playing football near their homes on patches of communal grass. These signs were photographed, depicted on maps and discussed within the groups. For example, Isabelle, age 15, explained of her photograph: "this is a sign that is on a piece of greenery on my road. It stops children from playing typical games, but little children need somewhere to play... they may not be allowed to go to the park". In discussion she said:

> "They've got 'No ball games' signs all over our streets, and there are loads of little pieces of grass where kids could just play, and be fairly happy and fairly safe, but they put up 'No ball games' signs and then they can't play there, and it's like stopping them from ... enjoying themselves."

Another girl, Katie, included the sign on her map, and wrote underneath "not fair".

Exclusion by Cost

The cost of activities was frequently mentioned as a barrier to participating fully in community life, and there appeared to be very few leisure opportunities for children in the town that did not involve spending money. A new leisure complex had recently been opened in the town centre, but all age groups in both sites complained about the cost. A group of younger children (12-13 year olds) had the following discussion:

> Cameron: "If there was a swimming pool, or a cheaper cinema, we'd be there."
> Dion: "If they made more ... like adventure centres, ... and things like that, then we'll be down there nearly every weekend, and things like that, but since they don't have anything like that we wander about the streets and get in trouble for it."
> Shannon: "The most fun thing we can do, without our parents having to pay for it, is just go and walk round the shops in town."
> Dion: "exactly."
> Angelina: "...and we can't afford it, 'cos we can't get jobs at the moment."
> Casey: "it's hard, 'cos me mum don't get paid that much."

In both schools the town centre was an attraction. Older children described how 'I like town better, there's much more action'.

However, children from West area (which was further away from the town centre than Hill area), complained that the bus fare rises to an adult fare when they are 13 or 14, but they don't have much money to spend, and as noted, the only entertainment is in town, and it is expensive. One girl in Hill area said: 'the thing that annoys me is that the police always moan that we're on the streets, so they build places like the new clubs and stuff, but we have to pay to get into that'. Further, when they did go into the town centre, they felt they were regarded with suspicion by security guards and shop-keepers who gave them 'dirty looks' because they thought they would be shoplifting (Matthews *et al.*, 2000; Morrow, 2000a; Woolley *et al.*, 1999).

Practical Exclusion and the Problem of Road Traffic

Road traffic was a major preoccupation, particularly for younger children. Often it was difficult for them simply to cross the road. Kellie described how in her neighbourhood, there was "nothing to do much, we live on a main road, it's quite busy, it's hard to cross the road". Rebecca added: "I don't like the person across the road, because we have to make sure when we cross the road, there's this man, that lives across the road, and he zooms round and he nearly hit my brother." Others complained about motorbikes joyriding, or riding on pavements, Kerry states:

> "We have motor bikes that come straight onto the mud track of the field, they go straight past the houses and down and you have little kids walking sort of like on the path and that, and they have motorbikes that go down there..."

In both schools, the issue of traffic led to long discussions and graphic descriptions of accidents and near-accidents:

> Cameron: "my brother got run over a couple of weeks ago, by a bus driver, and the bus driver just opened the window and swore at him, but ... [my brother] was on his bike, and

> he gave way, because they come so fast round the corner, he hit him, and just started shouting at him, through the thing, and left him in the road."
> Dion: "it's getting worse, like now, cos like when it was in my mum's, and your time, it would have been safer, out on the streets, now it's even worse."

The traffic was also a problem for Casey, who described how she baby sits twins and a little boy: "and I can't take 'em anywhere, because [the little boy] runs around the streets, and one time he was crossing the road he was about that much away from getting knocked over". A group of 12-13 year olds described how:

> Sonia: "There's traffic lights, you know down town there's that road, there was a car going by, there was a lot of traffic, the light was red, it just went past."
> Charles: "I was nearly run over outside school walking across the zebra crossing! A car speeded up and just missed me."
> Iftikhar: "When they're turning they should use the little yellow light [i.e. indicator] because they don't use it."

Traffic and busy roads can constitute practical constraints on children's environments and may impinge on their social interactions. Hillman's (1993) study showed how increased traffic was used by parents as a reason not to allow their children out on the streets on their own, and other recent research from elsewhere in UK has shown that traffic may constrain neighbourhood activities for children (Davis and Jones, 1996, 1997; Christensen and O'Brien, 2003).

Gender

Previous research has shown that girls tend to have more restricted access to public space due to parents' fears and stereotypical gender expectations (Valentine, 1997). Girls in the study were not confined to their homes, however, and described a great deal of time spent out in parks and on streets. However, as Matthews (2003) and Tucker (2003) have also found, girls appeared to be more fearful than boys, and accounts differed by gender, particularly among the older group: threats of sexual assault, anxiety about public spaces, were mostly (but not always) expressed by girls. This was unfortunately NOT a case of 'stranger danger', as is often claimed in the literature: at the time of the research, a serial rapist was attacking women and girls in the town - Natalie, aged 12, commented:

> "I don't feel safe where I live, because we've got flats near us, and because we've heard that people have actually been killed in those flats and stuff, and we have like rapists go round our area… it wasn't very long ago, and I don't exactly feel safe round my area."

And in an older age group, Amy described how:

> "Like someone was assaulted down [in the local park], I mean, that makes you scared to go down there, and that was in broad daylight, so God knows what its gonna be like at 10 O'clock at night. I live in like a secluded road, hardly anyone comes down my road, but there's nothing there, there's like a little park down the road, but someone was assaulted there, you're scared to go there. So if I was, like, 20, and I had two little kids, I'd have nowhere to take them in [this area], that was safe."

Girls were more likely than boys to show a concern for others, often expressing a strong sense of morality in their comments - and very often they showed a particularly concern for younger children, as mentioned above. In discussion, another girl described how:

> "Outside my house, we've got this green, and they put a

> notice up, saying 'no ball games', we had two trees, we used to use it as goal, there was this woman she always complained and she got the council to dig up the trees and put bushes on the grass so we couldn't play, we used to play rounders and stuff."

Another girl, Katie, age 13, included the sign on her map, and wrote underneath 'not fair'. The fact that girls highlighted this may reflect constraints on their mobility: they may want to play nearer to home and not in parks (where football dominated). As one 12 year old girl said:

> "I have to come in at 8 O'clock, but if I am in my road I can come in at 9 O'clock, something like that, but when I'm somewhere my mum don't know, kind of thing, when I'm out with my other friends, 8 O'clock."

It was mostly girls who described their parents looking out for them, one girl who had a newspaper round explained that her parents help her with it: "My mum and dad are overprotective, so they do it with me". Cameron, age 12, explained:

> "My mum don't like me going up [to another part of town] on me own, because you never know what its like, there's a rapist about at the moment up our area, so its hard. Because you have to watch where you're walking and you're not allowed to go anywhere by yourself, you have to be in a group, and if the police would actually do something about the rapist, then we could like go up [the Park]."

Cameron also complained about joy riding, and described how:

> "Round my area, there's these few boys, and they have this car, and they were speeding down the road, putting the brakes on, and swivelling round. I was walking my

> cousin home, and the car come up onto the pavement, and nearly knocked her over."

She suggested a solution: "most of the boys down our area are interested in cars, and motorbikes, if they could learn about mechanics, then maybe they would be off the streets". There was one example in a group discussion of two girls describing anti-social behaviour, Natalie was describing an incident, and admitted: "we were vandalising the bridge, I'm honest" at which point Agnes explained: "everyone vandalises our bridge".

Shortcuts, paths and routes away from busy roads were very important to children, but these were often experienced as frightening and unpleasant. One girl described how:

> "I hate walking through subways, I walk through two subways on the way to school, and I think, am I ever going to get to school? Cos you don't know whether there's someone hanging around the corner, or whatever, or following you behind, ... I hate walking through. Me and my sister just speed through them, but if I'm with like a bigger group, then I don't care."

Improving the Area

Children discussed ways that their neighbourhoods might be improved, and they made a range of practical suggestions. These can be clustered around the following themes: activities for their age group, and for older or younger children; suggestions about improving the traffic, and making the paths and shortcuts they used safer and cleaner; and finally, suggestions about the police.

Activities for Children

Most of the suggestions for improvement centred on the provision of leisure facilities. One girl said: "To make our neighbourhood better, I think they should do stuff for teenagers". Others said: "they should have more facilities", "make the parks better", "just somewhere to go". Other suggestions included better BMX tracks, more clubs, ballet and gymnastics and dance clubs, and so on. Bart, who lived in West area, described how 'Motorway Field':

> "is like a long strip, and at the end, there is this round bit, there's a few trees there, but it'd be nicer if ... they planted more trees there, so it was like a little mini forest where people can build dens, that won't be kicked in and stuff, so there's more variety of things to do."

As already noted, others expressed a strong view that if there were more facilities, they wouldn't undertake anti-social activities.

Paths, Routes, and Roads

As noted, some children in both study sites complained about their shortcuts and routes they took, and had ideas about how to improve them. A boy in West area described how "There's this great big wind tunnel, and you can't see around it, and you can't see who's in there, and there's often quite a lot of people in there that are like ... drunk". Children had a number of suggestions for making these routes safer and cleaner, including the installation of mirrors in subways, CCTV cameras, better street lighting, more bins for litter, and more signs telling dog owners to pick up dog mess. As we have already seen, road traffic was a described as a problem, and they had many suggestions as to how to slow down the traffic, including more traffic lights, parking on one side of the road to slow the cars down, more speed bumps and, not least, asking drivers to use their indicators.

Security, Safety and Fear

Many of the children in West area suggested putting more police on the street. As John said, "I've only seen policemen in cars, I've never seen a bobby on the beat". In Hill area, Jamel suggested "more police, probably". Wasef qualified this by saying: "but you don't want the police to be too hard, and you don't want them to be too soft". Jamel added: "there used to be one police officer around Hill area, but now you don't hardly see them". However, simply putting more policemen on the streets was not straightforwardly accepted by all children, and some young black men described how they felt under threat not only from other groups of young people, but also from the police. In Hill area, children felt as though they were specifically targeted by the police, as Asa May complained:

> "The police go round our area, looking for my age playing about, but they don't do anything about all the ones who have left school, driving around in their cars, they don't worry about them. And when they're looking for us, they could be doing other things, instead of looking for kids playing."

In summary, children were resourceful commentators on their environments, and they put forward many reasonable suggestions for improving their areas. However, one of the problems for this age group, as the next section shows, is that there are no consistent channels for them to communicate these views about their environments.

Exclusion from Participation

As noted, Putnam's (1993) emphasis on civic participation as a key aspect of social capital is obviously somewhat limited in the case of children, given that they are positioned outside of democratic structures by their very nature as 'minors'. Indeed, participation in community decision-making for children in the study was extremely limited. Only one boy felt he could go to his local residents' association and make suggestions about his local area. If the council did come and ask about local facilities, they felt that their parents were consulted, not them. Amy said: "they send questionnaires to our parents but it's not our parents who want to go to the Youth Club, it's us. So they should ask us." One girl said "I don't think people are really bothered about kids"; in another group a boy said: "They just do things like little tiny parks for little kids... we don't want little parks".

One girl commented that she felt they should have a say in the community, "because what happens does affect us as well as the adults and they don't seem to think about that when they're making decisions." The town council has recently started a 'Youth Forum', but children in the study were not aware of it:

> Gemma: "No-one knows about it, if there is one."
> Tamisha: "I think there should be one, but..."
> Miranda: "...but they'd chose the people who do all the best in school, and everything, and they're not average people, are they?"

These data suggest that participation, in the sense of being actively involved in decisions that affect them in their neighbourhoods, appears to be virtually non-existent for these children. Even where supposedly democratic structures such as school councils are in place, as was the case in one of the schools in the study, children did not seem to feel they were experiencing 'participation' through them, and the exclusion they appear to feel is likely to limit their sense of self-efficacy and control over their environments (Morrow, 2000b). One of the problems facing this age group is that they have no formal channels through which to communicate, or to convert their energy into a positive resource for their neighbourhoods. Youth forums are the most common way of facilitating children's views, but they do not necessarily work effectively (Fitzpatrick *et al.*, 1998). Miranda's comment, above, suggests that she is well aware of the limits of democratic participation and representation.

Problems and Possibilities

In summary, the paper has suggested that children's experiences of their physical environments were mostly negative. In 'social capital' terms, they did not appear to derive a sense of belonging from an identity with their town. The overall study found that social interaction with friends was, from the children's point of view, crucial for a sense of belonging. However, they described a number of features of urban space that impinged on these social interactions, such as traffic, and lack of facilities. Hill Area was perceived as particularly inadequate because of the lack of green spaces to play ball games in. The town centre was a strong attraction, but often experienced negatively because of the 'dirty looks' they got from security guards, shopkeepers, and other adults. Further, visits to the town centre involved spending money, or the temptation to spend money, and cash was often limited.

In terms of possibilities for improving their neighbourhoods, as we have seen, children put forward and discussed many reasonable and practical suggestions. However, they were well aware that they had limited self-efficacy and participation in their neighbourhoods. As noted, in terms of civic participation, this is not really surprising, but the experiences they have *as children* may have implications for their perceptions of democratic institutions and structures later on when they do leave school, and this begs the question of whether or not a 'healthy scepticism' is learnt early on in life. They were frequently faced with situations that did not meet their perceived needs, and sometimes the only forms of resistance available to them are 'anti-social' ones. Young people and children are rarely regarded as 'stakeholders' in their communities and their perspectives and views are not consistently elicited, as the data reported in this paper show (see also Speak 2000).

Putnam's (1993) notion of social capital hinges upon being able to define 'community' as a geographically circumscribed area, and for children this is problematic, because, as we have seen, 'community', if it exists at all, appears to be located in a sense of 'belonging' that resides in relationships with other people, rather than places. The study suggested the following elements of social life comprise 'community' for children: school is a 'community' in that it is an important site for social interaction; their neighbourhood may be experienced as a safe or hostile community;

and they may be members of a community of interest, such as a dance class located in a neighbouring area or a sports club in a different part of town. For children, then, 'community' is more often a 'virtual' community of friends based around school, town centre and street, friends' and relatives' houses, and sometimes having two homes, and thus two sets of friends in two different towns, rather than a tightly-bound easily-identifiable geographical location.

Urban spaces are likely to be experienced differently according to gender, age and cultural or ethnic background. The study highlighted how 'children' do not constitute a homogeneous category. 'Social capital' needs to be able to accommodate a range of different social identities. For girls, personal safety was a crucial issue, and sexual assault was perceived as a threat. There were differences in children's accounts according to age: the needs of 12 year olds, for example, for places to play and make dens, differed from the needs of 15 year olds, for example for places to socialise away from the sometimes hostile gaze of adults. Further, there were significant differences in priorities according to ethnic background, and racial harassment was described by several children. Such harassment is likely to lead to social and emotional exclusion, and this kind of hostility is likely to inhibit social cohesion at the neighbourhood level (see also O'Brien *et al.*, 2000; O'Brien, 2003).

As we've seen, a range of practical, environmental and economic constraints were felt by children. The extent to which they move around freely to participate in activities with their friends is likely to be constrained by the physical geography of the built environment, issues of community safety and traffic, parental norms about when children may go out, as well as access to financial resources. These issues are usually neglected in studies of children's health behaviours. Perhaps these constraints might be more usefully conceptualised as influencing access to 'social resources', rather than 'social capital' *per se*. There is a danger that in operationalising social capital in research based on Putnam's (1993) conceptualisation, other forms of capital (material, economic, symbolic, spatial) are overlooked, as the 'social' is reified and separated out as a unit of analysis.

There are undoubtedly advantages to using social capital as a research tool, and the 'social capital' research in the UK has undoubtedly enabled social policy research to focus on the circumstances of people's everyday lives. In the case reported here, a focus on 'social capital' as a community level attribute has allowed research to prioritise the social context of children's everyday lives, rather than their individual health behaviours. Children's subjective views and explanations of their social worlds, their neighbourhoods and networks, have been highlighted, and these offer perspectives that differ from adult-oriented preoccupations about the needs of this age group. These data show clearly that they have views that they are well able to articulate about their social environments. It is arguably necessary to try to understand whether children have a sense of belonging in their neighbourhoods if we want to try to bring children's views into the policy debates around public health (Davis and Jones, 1996, 1997). 'Social capital' is thus useful as a tool for exploring social processes and practices around children's experiences of their environments, and in doing so, it has highlighted children's 'social resources' (or lack of them). The implication of this is that 'social capital' needs to be conceptualised in relation to the other forms of capital that underpin it or are related to it. Space, it could be argued, is an important form of capital in that it represents a resource that impacts upon social relationships and social networks - but space is also at a premium in UK towns and cities, and children are not in a position to make claims on it. Exploring these issues in relation to children and young people has focused on one element of social life, that is, neighbourhood quality of life and use of urban space. There is a danger that in doing so, broader questions may be overlooked, and the wider political responsibilities of central and local governments to ensure a good quality of life for children in environmental terms also need to be addressed.

Acknowledgements

The author is immensely grateful to the young people who participated in the research, and to the staff of the two schools, and representatives of the local Health Promotion Agency and Town

Council for their help in setting up the research. The research was conducted under a research fellowship funded jointly by the Health Education Authority and the Eleanor Rathbone Trust. *A longer version of this paper appears as a chapter in Christensen and O'Brien (2003).*

NOTES

1. The research did not explore the direct effects of parents/family social capital on children (see, for example Allatt 1993, 1996). Rather the intention was to explore the advantages and disadvantages of conceptualising social capital as a community level attribute in relation to children.
2. see Morrow (2001b) for a full discussion of methodological and ethical considerations.

REFERENCES

Adams, R. 1995. "Places of childhood" (pp.157-168) in P. Henderson (ed.), *Children and Communities*. London: Pluto Press.

Allatt, P. 1993. "Becoming privileged: the role of family process", (pp.139-159) in I. Bates and G. Riseborough (eds.), *Youth and Inequality*. Buckingham: Open University Press.

Allatt, P. 1996. "Consuming schooling: choice, commodity, gift and systems of exchange", (pp.163-182) in S. Edgell, K. Hetherington, and A. Warde (eds.), *Consumption matters: The production and experiences of consumption*. Oxford: Blackwell/The Sociological Review.

Backett, K. 1992. "Taboos and excesses: lay health moralities in middle class families", *Sociology of Health and Illness*, 14(2): 255-274.

Boyden, J. and Ennew, J. 1997. *Children in focus: a manual for participatory research with children*. Stockholm, Sweden: Radda Barnen.

Brannen, J., Dodd, K., Oakley, A., and Storey, P. 1994. *Young people, health and family life*. Milton Keynes: Open University Press.

Christensen, P. and O'Brien, M. (eds.), 2003. *Children in the city: Home, neighbourhood and community*. London: Routledge Falmer.

Davis, A. and Jones, L. 1997. "Whose neighbourhood? Whose quality of life? Developing a new agenda for children's health in urban settings", *Health Education Journal*, 56: 350-363.

Davis, A. and Jones, L. 1996. "Children in the urban environment: an issue for the new public health agenda", *Health and Place*, 2(2): 107-113.

Fitzpatrick, S., Hastings, A. and Kintrea, K. 1998. *Including young people in urban regeneration. A lot to learn?* Bristol: The Policy Press.

Henderson, P. (ed.). 1995. *Children and communities*. London: Pluto Press.

Hillman, M. (ed.). 1993. *Children, transport and the quality of life*. London: Policy Studies Institute.

James, A. and Prout, A. (eds.). 1997 (2nd ed). *Constructing and reconstructing childhood*. London: Falmer Press.

Matthews, H. 2003. "The street as liminal space: the barbed spaces of childhood" (pp.101-117) in P. Christensen, and M. O'Brien (eds.), *Children in the city: Home, neighbourhood and community*. London: Routledge Falmer.

Matthews, H., Taylor, M., Percy-Smith, B., and Limb, M. 2000. "The unacceptable *flaneur*: the shopping mall as a teenage hangout", *Childhood*, 7(3): 279-294.

Mayall, B. 1994. *Negotiating health. Primary school children at home and school.* London: Cassell.

Michell, L. 1997. "Pressure groups: young people's accounts of peer pressure to smoke", *Social Sciences in Health*, 3(1): 3-17.

Morrow, V. 1999. "Conceptualising social capital in relation to the well-being of children and young people: a critical review". *The Sociological Review*, 47(4): 744-765.

Morrow, V. 2000a. " "Dirty looks" and "trampy places in young people's accounts of community and neighbourhood: implications for health inequalities"," *Critical Public Health*, 10(2): 141-152.

Morrow, V. 2000b. " "We get played for fools": young people's accounts of community and institutional participation" (pp. 143-157) in J. Bull and H. Ryan (eds.), *Changing families, Changing communities: researching health and well-being among children and young people*. London: Health Education Authority.

Morrow, V. 2001a. "Networks and neighbourhoods: children's and young people's perspectives". *Social capital for health series*. London: Health Development Agency.

Morrow, V. 2001b. "Using visual methods to elicit young people's perspectives on their environments: implications for community health promotion initiatives", *Health Education Research: Theory and Practice*, 16(3): 255-268.

Morrow, V. 2001c. "Young people's explanations and experiences of social exclusion: retrieving Bourdieu's concept of social capital", *International Journal of Sociology and Social Policy*, 21(4/5/6), 37-63.

O'Brien, M. 2003. "Regenerating children's neighbourhoods: what do children want?", (pp. 142-161) in P. Christensen, and M. O'Brien (eds.), *Children in the city: Home, neighbourhood and community*. London: Routledge Falmer.

O'Brien, M., Jones, D., Sloan, D. and Rustin, M. 2000. "Children's independent spatial mobility in the urban public realm", *Childhood*, 7(3): 257-279.

Pavis, S., Cunningham-Burley, S. and Amos, A. 1996. "Young people and smoking: exploring meaning and social context", *Social Sciences in Health*, 2(4): 226-243.

Pavis, S., Cunningham-Burley, S. and Amos, A. 1997. "Alcohol consumption and young people: exploring meaning and social context", *Health Education Research: Theory and Practice*, 12(3): 311-322.

Putnam, R.D. 1993. *Making democracy work: Civic traditions in modern Italy.* Princeton N.J.: Princeton University Press.

S.E.U. Social Exclusion Unit, Cabinet Office. 2000. *National Strategy for Neighbourhood Renewal: Consultation Paper.* London: HMSO.

Speak, S. 2000. "Children in urban regeneration: foundations for sustainable participation", *Community Development Journal*, 35(1): 31-40.

Tucker, F. 2003. "Sameness or difference? Exploring girls' use of recreational spaces", *Children's geographies,* 1(1): 11-124.

Valentine, G. 1997. " "Oh yes I can". "Oh no you can't": children and parents' understandings of kids' competence to negotiate public space safely", *Antipode*, 29(1): 65-89.

Woolley, H., Spencer, C., Dunn, J., Short, T. and Rowley, G. 1999. "The child as citizen: experiences of British town and city centres", *Journal of Urban Design*, 4(3): 255-282.

Chapter 4

BASKETBALL IS LIKE BREATHING

Young People's use of Urban Physical Activity Spaces

Matthew Atencio

Historically an economically disadvantaged area, the inner North East Portland neighbourhoods were home to frequent 'gang banging' and drug dealing during the mid-1980s to mid-1990s. As one local writer put it:

> "In the mid-1980s came King Crack, provider of income, sleek cars, and clothes and an unparalleled high. It was a jobs program for North East Portland the same way it was for South Central Los Angeles and East Oakland. In those days, Woodlawn Park Crips and Kerby Street Bloods set handguns barking against the night…" (Dawdy, 2000: n.p.).

This recent dubious past still weighs heavy on the minds of local residents. As one African-American resident put it, "It's hard to get that [past] out of our minds. It's [the North East neighbourhoods] still the so-called "hood" to us" (Suo, 2001: n.p.). Although gang activity has declined since 1997, there has been a recent resurgence of gang-related shootings in North East Portland (15 shootings in 2005) and the 'Crips', 'Bloods', and Latino gangs still remain active in the area (Beaven, 2005).

In early 2003, I moved into the inner North East neighbourhoods of Portland. Through the course of my observations it became evident that basketball is a very popular and high-profile sport in this area. Well-supported basketball camps, open gym programs community tournaments, and high school and club teams exist in the local landscape. I've seen young children sitting on front porches tossing tiny balls into the encircled arms of mothers on summer days. Even a homeless man walks down the main thoroughfare bouncing a basketball. Turn down many streets and young people can often be seen throwing balls up towards portable or makeshift basketball hoops (see Figure 4.1). Park courts refurbished by recycled Nike shoes are filled to the brim, especially during the summer on weekend afternoons when these spaces play host to numerous pick-up games and community tournaments (see Figure 4.2).

In my study a group of 16 young men and women was recruited through the local high school and community centres. This group comprised of young people from diverse racial, ethnic, and national origins, and included seven young women and nine young men. 13 of the 16 young people attend or graduated from Thompson

Fig. 4.1 : A typical portable or makeshift basketball hoop.

Fig. 4.2 : An example of a local community basketball tournament.

High School which has a high percentage of 'ethnic minority' enrolment.[1] These young people participate in sports and physical activities such as basketball, skateboarding, dancing, running, and football.

Qualitative methods such as open-ended individual and group interviews, photography, videotaping, and observations were used to collect data. The young people were aged between 13 and 18 when they first became involved in the project. Over the course of three years I conducted seven interviews with each person. The interview topics were related to health, physical activity, popular culture, and neighbourhood spaces. Issues of gender, race, social class, family life, and peer relationships were interwoven throughout these interviews.

This chapter draws from four interviews conducted per person describing their usage of urban spaces while participating in sport and physical activities such as basketball. In these interviews the young people were asked to describe which local spaces they used during one week in the summer and one week during the school session. Young people were also asked to outline their reasons for choosing certain spaces over others and to describe their experiences within their preferred spaces. City maps featuring parks, schools, streets, and playing fields were used during interviews to help young people reconstruct their physical activity movements. Descriptions of young people's movements were transformed into Microsoft Excel spreadsheets and then into multi-dimensional Geographic Information System (GIS) maps that were developed through the resources of the Portland State University Department of Geography[2].

In this chapter, I take up the notion that spaces are resources which young people work with to understand who they are and to interact with those around them. Before offering descriptive accounts and analysis from the spatial interviews I will first provide a relevant literature review around space, identity, and culture.

Rethinking Physical Activity Spaces and Identities

This chapter draws from several intellectual geographic traditions in order to describe the impacts of social space in the lives of physically active young people. Urban, environmental, and cultural geography analyses have been used to investigate the complex relationships between society, people, and spaces. This has been achieved by drawing from Foucault's (1973) concept of discourses. Green (1998: 183) describes discourses as :

> structured ways of knowing which are both produced in, and shape culture. They are always laden with power and institutionalized as practices, which in turn help shape individuals as subjects.

In this context, individuals are not 'passive recipients of constraining structures' but instead are 'active agents' negotiating various intersecting and contradictory discourses in order to construct a sense of who they are (Green, 1998: 172). By drawing on this concept of discourse and subjectivity, these geographic approaches have examined the social processes by which spaces and identities are constructed in relation to gendered, racialised, and class discourses. It has been well established in these geographic academic fields that spaces provide individuals with the discursive resources from which they can form various identities. However, the disciplines of sport and leisure studies have been slow to take up these points of analyses (Aitchison, 1999). This suggests the need to further examine how sport and physical activity spaces work to provide individuals with particular classed, gendered, and racialised ways of thinking about their selves and bodies (Crouch, 2000).

Many theorists have argued that we live in a postmodern world where people engage with time and space in a non-linear, fluid, and 'hyperreal' manner (Baudrillard, 1988). In this context young people negotiate urban spaces in ways that are linked to local and global social conditions and symbolic cultural representations. It has been noted that the new post-industrial economies are 'more concerned with the production of symbolic values than the production of material commodities' (Andrews, 1996a: 316). In this context "our material existences become negotiated through and against the proliferating outpourings of the *new* culture industries" (Andrews, 1996a: 316). It is, therefore, important to analyze the impacts of post-industrial economies in the lives of young people. There is a need to examine the ways individuals draw from and make choices about cultural and symbolic representations in order to assume various forms of subjectivity and to negotiate the physical activity spaces around them. For example, an important point of analysis in this chapter will be to examine the ways the Nike swoosh has come to symbolize individual empowerment and civic responsibility through the mass marketing of *mantras* such as 'just do it' and 'self-made' athletes such as Michael Jordan (Cole, 1996). This raises the question: What meanings and identities are taken up when young people use basketball courts laden with Nike swoosh symbols, wear Nike-sponsored clothing and shoes, and watch Nike television commercials[3]? Accordingly, this study examines the practices through which young people activate, influence, and negotiate various spaces, social contexts, and cultural representations simultaneously (Crouch, 2000).

A focal point of this analysis is then to describe how young people engage with the identities and spaces that have been constructed by the new symbolic cultural industries as the racialised 'Other'. Aitchison (1999: 31) argues that the 'othering' process "inevitably defines norms and deviants, centres and margins, cores and peripheries, the powerful and the powerless, and is characterized by the post-Enlightenment binary divides and dichotomies". Rose (1995, as cited in Aitchison, 1999: 31) notes that through the process of constructing and contrasting oneself with the 'Other', individuals are provided with the means to define who they are and where they belong. By constructing an 'Other', particular knowledges and identities are legitimised as 'normal' and 'appropriate'.

Andrews (1996b) for example, has examined popular cultural representations of a naturally athletic African American or Black 'Other'. Specifically, he argues that popular media stereotypes have positioned black urban youth as being both pathologically criminal and naturally athletic. These racial discourses are linked to widely held assumptions about the 'innate physicality' (1996b: 135) of black bodies, which can be traced back to the slavery era. It has been argued by Hokuwhitu (2003: 30) that this construction of an African American 'Other' serves to reify hopeless sport dreams and "fixate people of colour into roles acceptable to a social narrative of normalcy". At the same time, the othering process works to legitimise the role of Whites to be "safe keepers of civilised intelligence" (Hokuwhitu, 2003: 29) and providers of "enlightened leadership" (Hokuwhitu, 2003: 28).

This construction of a racialised urban 'Other' has been used to justify interventionist sport and physical activity programs in poor urban areas (Hellison, 1995 and Martinek & Hellison, 1997). Private and public sponsored sport programs such as Midnight Basketball "target underserved youth, who are often referred to as disadvantaged, at-risk, Black, and inner-city youth..." (Pitter and Andrews, 1997: 86). This follows a widely held belief in sport to "properly socialize citizens to appropriate civic engagement, human interaction and, of course, productivity" (Jamieson, 2003: 6). However, basketball spaces have been implicated by several researchers as reifying racial and class inequalities and simultaneously (re)producing an Anglo-centric middle class *ethos* of social intervention and individual empowerment (Pitter and Andrews, 1997: 93). Under the pretext of social intervention, sport spaces are used "to regulate, discipline, and police already deviant bodies in urban areas" (Cole, 1996: 388). Taken together, these researchers provide compelling evidence that powerful social imperatives are at work in urban physical activity spaces.

This chapter engages with these White middle-class constructions of an at-risk urban Black or African American 'Other' and presents evidence that this positioning directly affects the ways these young people use their neighbourhood spaces. By engaging with these spaces 'individuals comes to know about the world in new, more complex ways' (Crouch, 2000: 65). It follows that young people construct their identities based on the knowledges produced by their encounters with physical activity spaces. By taking up particular identities in relation to the physical activity spaces around them individuals come to make judgments about themselves and others. Green (1998) argues that through their encounters with spaces individuals take up particular identities from which they form inter-personal relationships and construct social hierarchies. Following on, class, gender, and racial identities are integrally linked to the operation of social hierarchies in urban spaces. These hierarchies serve to exclude and constrain young people and (re)produce notions of 'acceptable' identities and behaviours (Aitchison, 1999, 2001; Mowl and Towner, 1995; Scraton, 1994; Scraton and Watson, 1998; Shogan, 2002).

By drawing on the literature above, the focal point of my analysis will be to investigate how basketball spaces work to maintain and (re)produce complex power relationships. Accordingly, this chapter describes how the operation of social hierarchies in basketball spaces is integrally linked to individuals' performances of race, gender, class, and physical capability.

Young Men's use of Basketball Spaces

Adam and Neal[4] are two African-American young men who were 15 years old when they entered the study- both are now 18 and preparing to accept university athletic scholarships to play basketball next year. They play basketball for the state championship contending Thompson High School[5] varsity team and for elite club programs in the summer time, programs which travel to prestigious tournaments across the United States.

In the following analysis I will draw from their 'summer' physical activity maps. On a typical weekday, Adam and Neal get out of bed around 12:30 pm and then go to a friend's house to hang out (Adam often spends the night at Neal's house). They sometimes play video games, talk to people on the phone, and watch television. Other times they will walk across the street to Peninsula Park to shoot the basketball around. In the following exchange Adam and Neal describe the scene at Peninsula Park:

> INT : *Um, and then you hang out until about 4 o'clock, play video games, and then which park, say it's Monday, where would you go to?*
> N : *Like at first, most of the time we'll go across the street and shoot around for a little bit...*
> INT : *And where is that?*
> N : *At Peninsula, for a little bit.*
> INT : *So at Peninsula, where the (outdoor) court's at...*
> N : *Yeah... And then, we'll all ride our bikes over to... Irvington.*
> INT : *Okay, so at Peninsula you shoot around a little bit,*
> N : *Yeah...*
> INT : *And, uh, how come you don't play at Peninsula?*
> N : *'Cause, nobody be there...nobody be there... and people that be there be weak...we want to play some competition.*

Friends will later call them (their friends are usually team mates) to say that Irvington Park (actually named Irving but often referred to as 'Irvington') is 'crackin'[6] that afternoon. The news spreads through this informal phone network and Adam, Neal, and their friends will ride their bikes over to Irving Park, which is considered the hotspot for park basketball players in Portland: By 4:30 pm these parks will have 'two courts full' of people playing basketball with large crowds around the courts.

From Adam and Neal:

INT : *So where are the best crowds at, do you think?*
N : *Irvington [Irving Park]...*
INT : *Um, is the crowd important to you or not?*
A : *Yeah, I like having a crowd better because without a crowd or a little bit of people there at least, I mean, it's not as fun, it's fun still, but it's not as fun, 'cause you have like people encouraging you and stuff to do good and stuff and then you have if you make a good play or somethin' the crowd goes 'ooohhhh' and stuff and you like that. So it gets you like, kinda like the crowd gets you motivated too, it's better.*

Irving Park is preferred by Adam and Neal over other places like Alberta Park because of the higher level of competition and the ability to perform better in front of a crowd.

From Adam and Neal:

INT : *Okay, Adam you were saying you like Irvington better (than other parks) - how come?*
A : *I like Irvington better because there's more competition and just older people up there. So, it's like, more competition. I like playing against older people.*
INT : *Why do you like playing against older people? They're bigger...?*
A : *They're bigger and stronger, there's more competition.*

INT : Uh, huh...
N : We be killin' people our age.

Because the majority of players at Alberta Park are males ('it's mostly dudes') who are the same age or younger, there is not enough competition for Adam and Neal, who dominate players in their own age groups. For this reason they choose to play elsewhere. On Fridays, after playing at Irving Park until 7:30 pm these young men will ride their bikes to a friend's house and hang out. Then around 8:30 pm they will leave their bikes and walk over to the nearby Boys and Girls Club. They shoot the basketball around in the gymnasium with mostly younger kids, and then at midnight they will go over to the Midnight Basketball[7] program at the Salvation Army community centre (known as the 'Sal'). The level of competition is high because many older and more physically developed players are there. For Adam and Neal spaces like Irving Park and the 'Sal' provide the highest level of competition in the area and the chance to play 'harder' in front of the best crowds.

From Neal:

N : If there's a lot of people at the park then, you want to play good. You just want to do good, so you just have everybody like 'oohhh' and do like that's when you, that's when I want to just like dunk more and try to do different stuff.
INT : Uh, huh...
N : Play your hardest, when there's a lot of people I play harder. And like at Salvation Army I got to play my hardest 'cause it's like stronger people and like, they like sometimes, mostly they better than me, but, you know I love showing you know, I just play hard, you know, and try to do my best.

Adam and Neal will play in the Midnight Basketball program at the Salvation Army until it closes down at 2 am and then they will walk back to their friend's house and then ride their bikes back home. In the previous exchanges Adam and Neal described competition levels in various northeast parks as being a major factor in choosing where to play basketball. While Adam and Neal often play in the parks, particularly in the summer, they are primarily concerned with winning ('running the court') and getting themselves and their team mates (they usually play with their team mates in the park) ready for upcoming organised or institutional competitions.

From Neal:

INT : How do you feel about park basketball, street basketball, compared to say the high school team or your club team?

N : Street basketball to me, is better, for me, it helps me out with my high school basketball 'cause I'm tryin' to get stronger people (to play against), and you know, better people, quicker, you know they got more handles and jumping so, I gotta play extra hard and extra strong 'cause you can't really call a foul so you gotta just play up, so, you know, I work on my shot, when I get fouled if the referee don't call it, I still knock the shot down, you know, I still keep playing without calls from the referee and that helps me in my high school that's helpin' me with my high school, you know, so I, I won't be lookin' for the ref to bail me out I just keep playin'.

The codes and rules of park basketball (less foul calling and winning teams stay on the court) provides top level players such as Adam and Neal with the opportunity to improve their basketball skills and to prepare for their high school and club seasons. Because Adam and Neal are very capable players and choose to play with their highly-talented high school team mates, they are often able to 'run' the courts in a variety of park spaces.

'I got "next" '

In the park basketball spaces, players who show up to the court call out 'next' or 'down' to reserve their spot in a future game.

Whoever calls this waits on the sideline until it is his or her turn and then eventually gets to pick a team to play the winners for the right to stay on the court. Players whose abilities are perceived to be inadequate will be unable to play since they will not get picked to play on a team. In this way many young people are excluded from using park basketball spaces[8]. In the next quote Ricky, a 16 year old African-American male who also played for Thompson High School, describes how 'next' is directly related to perceived physical capabilities rather than race:

> INT : *So a lot people from all over go up there [to Irving Park]?*
> R :L*Yeah, all kinds of people go up there, all different kinds of races.*
> INT : *Tell me about that.*
> R : *Everyone goes to Irvington for competition.*
> INT : *So is it like blacks against whites?*
> R : *No, no it's not like that, it's just whoever gets picked, like they always have to have a big guy, and a small point guard, like at Peninsula, I'm always on the court, cause they know me and I'm they know I'm fast, at Irvington, you might get two games in, cause there's always people up there.*
> INT : *So are there some people that don't play?*
> R : *It might be, they might go up there to play, but they'll be watching all day.*

In 'informal' spaces such as Irving Park, where competition is high and places on teams are at a premium, the ability to perform on the basketball court provides access. Ricky's statements provide evidence that there is a hierarchy inside and between Irving and Peninsula parks. Ricky will always get picked to play at Peninsula Park because he is fast and athletic and is, therefore, a mainstay on the court. Irving Park is more difficult (he 'might get two games in') because it is a more competitive space than Peninsula Park. While Ricky was also a member of the prestigious Thompson High School basketball team with Adam and Neal, he wasn't a consistent starting player like they were. Adam and Neal, who are really 'known' in this area for their basketball-playing prowess, are able to play at Irving Park for much longer than Ricky.

John

John was 15 years old when he became involved in the study. John lives close to Woodlawn Park and dreams of playing in the National Football League. John is much shorter than Adam, Neal, and Ricky and has a stockier built body. He does not play for the Thompson High School basketball team but instead plays for the junior varsity football team. During the summer he also enjoys playing basketball with his friends. In the following excerpt John talks about avoiding Peninsula Park and Alberta Park and how he would prefers to play basketball in Woodlawn Park:

From John:

INT : *What do you think about Peninsula [Park]?*
J : *It's cool, but it's not like Woodlawn [Park].*
INT : *What do you mean?*
J : *My peers are at Woodlawn, it's a different crowd there.*
INT : *So you're unfamiliar with those people, how does that make you feel?*
J : *You can't talk trash that much 'cause' someone might pull out a 'heater' or something.*
INT : *Say that again...*
J : *Someone might pull out a gun or something.*
INT : *Has that ever happened to you?*
J : *Someone shot at my friends.*
INT : *Where did that happen?*
J : *Alberta [Park].*
INT : *Tell me about that.*
J : *My homeboy, he's kind of big, he was posting up, he turned around and slammed on a dude, and fell on top of him, and was like 'Aaaahhhh you can't let me do that man' and the guy was like 'whatever, do that again and I'll smoke you' and he told him he was going to smoke him.*

INT : *Does that influence why you don't go to Alberta?*
J : *A little bit, gangstas are at Alberta all the time, trouble over there.*

John avoids going to Peninsula Park and Alberta Park[9] because he doesn't know the people in these spaces and he is worried about his safety. This is a much different scenario compared to established basketball players such as Adam, Neal, and Ricky. These types of confrontations and safety issues were not found in the interviews with these players. This is evident in the following exchange with Neal:

INT : *Tell about that, they [the gangsters] don't mess with you because it's your life, because you play basketball.*
N : *Yeah, 'cause like, we play in the park and we'll play against 'em and we'll, we'll be killin' them, and they be like 'oh, he good' and then they talk to us and tell us like stories about them and how they used to be good and stuff that like that so we get to know 'em better. And then like, if we have problems with somebody then they just be like 'don't mess with him' and stuff like that so we really don't got no enemies by hoopin'.*

Because Adam and Neal are known as good players they are not harassed by gang members and they are actually supported by them - one north-east precinct Portland police officer told me that many of the gang members in these parks want to be 'ballers' (good basketball players) too. While John is constrained from using Alberta and Peninsula parks because of the gang members, Adam and Neal wouldn't even want to play in these parks because of the lack of competition.

From Adam and Neal:

INT : *Okay, Adam, you were saying you like Irvington better (than Alberta Park) - how come?*
A : *I like Irvington better because there's more competition*

and just older people up there. So, it's like, more competition. I like playing against older people.

INT : *Why do you like playing against older people? They're bigger...?*

A : *They're bigger and stronger, there's more competition.*

INT : *Uh, huh...*

N : *We be killin' people our age (at Alberta Park).*

The previous excerpts demonstrate how the young men in North East Portland are all positioned in different ways in relation to the hierarchies of North East Portland park basketball spaces. The parks serve as a training ground for people like Adam, Neal, and Ricky to improve their basketball abilities - these spaces are an extension of elite boy's club and high school teams. Because these young men have high levels of ability they are able to maintain a presence in all of the park court spaces. They are also able to play in all the park spaces without fear because they are talented basketball players and, as Neal says, 'we got no enemies by hoopin'. The 'gangstas', because they have the capacity to chase other young people off the courts, are able to play in spaces such as Peninsula and Alberta parks. However, they are unable to 'run' more competitive spaces such as Irving Park which requires higher levels of skills for access. Players such as Adam and Neal position all the 'other' males, including the gang members, as getting 'killed' or as being 'weak' because they aren't as talented as they are.

In Irving Park, John wouldn't be picked to play in a game because he isn't really 'known' in the area as a basketball player. In Alberta Park he was chased away by someone who threatened him with a gun. John can only play basketball at Woodlawn Park close to his home - he says that 'It [Woodlawn Park] is so convenient and like all my friends are there, everybody knows me there. So it's a friendly place to be'. In relation to the hierarchies of North East Portland basketball spaces John isn't even positioned as being 'weak' or getting 'killed' because he's not even a presence in the majority of the parks. He's essentially invisible in these basketball spaces and is limited to using Woodlawn Park next to his house.

Young Women's use of Basketball Spaces

Despite his marginalized status in relation to the park basketball spaces of North East Portland, John talks about how he excludes and 'plays rough' with the young women in Woodlawn Park:

INT: *What about when the girls play, what is that like?*
J: *We have fun.*
INT: *How come?*
J: *Because they would be always talking about don't foul them and [we] plays rough with the girls.*
INT: *Do guys let the girls play or do the girls play, do they want to play?*
J: *Oh yeah the girls want to play sometimes but they don't let, some people don't let the girls play.*
INT: *How come?*
J: *Because the girls that go down to Woodlawn aren't as good as the dudes so they don't want to get, the dudes don't want to pick them up.*
INT: *Like when you play 'next' or something like that?*
J: *Yeah.*

Later in the interview:

INT: *So what if a girl comes down the park, then what?*
J: *She wants to play; she can play but she might not get picked up if she is good or not.*
INT: *Does that happen sometimes?*
J: *Yeah.*
INT: *What do you mean?*
J: *Like growing...like picking all my team.*
INT: *So what do the girls do then?*
J: *The girls go to a different court like a different basketball hoop.*
INT: *And do they play or do they shoot around or what do they do?*
J: *Just shoot around I guess.*

This excerpt from John demonstrates how young men employ tactics such as playing rough and not picking the young women to play by using 'next' to keep them off the courts. In many of the park basketball spaces young women are excluded or constrained from being on the courts while the young men are playing. Women often have to play on less desirable courts (temperature is a factor especially in the summer) or move to different spaces such as indoor gyms in order to play. In contrast to the young men this exclusion takes place even if they are very talented. This is evident in the next quote from Sharon, an African-American/Brazilian young woman who was 16 when she joined my study:

From Sharon:

INT : *Why would you go to Irving Park?*
S : *I don't know, it doesn't have so many people, not so many crazy[10] people like at Peninsula Park.*
INT : *What do you mean?*
S : *Black people hang out there, and they shoot and fight all the time, and its people my age, and they're fighting out there, and during the day the boys have the court.*
INT : *Do they hog the court?*
S : *Yeah, and they fight a lot, so they'll be playing a basketball game and someone will get mad and they'll start a fight, but at Irving Park, a lot of older people will go there and so its not like they're as horny as the other boys.*
INT : *And do you play with the boys or the girls?*
S : *The girls play there, but when you have the boys play there, they'll rarely pick you, or if they pick you, they won't really play with you, like they won't pass the ball, and be like 'she can't do it'.*

Despite the fact that Sharon is a very capable runner and basketball player she is unable to play in both Peninsula Park and Irving Park. Peninsula Park is considered to be unsafe and the 'boys have the court'. When she goes to Irving Park she is unable to get into a game because the young men won't pick her - if she does manage to get on the court they won't pass the ball to her. Sharon's comments echo the sentiments expressed by the young women in Wilson *et al.'s* (2001) study of an inner-city recreation centre - in their 'informal' basketball spaces these young women were only conditionally accepted into the male-dominated games.

Tara, an African-American young woman who is also highly talented at both track running and basketball and was 15 when she joined the study, describes similar feelings about being excluded from park basketball courts:

INT : *What about like I've talked to some of the people about who plays in the courts and things like that. What cultural groups are playing there and who plays, boys or girls or who gets to play?*

T : *Well guys tend to not really, well some guys, not all guys but guys tend to not really want to play with girls because they are girls and they feel that girls probably aren't strong enough or whatever. But like if the girls are locally known like they are know to play basketball, then the boys really don't check. But most of the time it's just like a group of guys there like the girls really don't tend to go to those courts as much as the guys do, because it will just be a whole bunch of guys and so the girls just feel kind of outnumbered and so they just kind of go somewhere else.*

INT : *Why do you go somewhere else?*

T : *Well because mainly at the main parks there the courts will usually be taken up, you know, so we just go somewhere else or we just really don't like being outside, just inside gym is better.*

INT : *So where would you play indoors then.*

T : *Sometimes when 'Thom' [nickname of the high school] is open we come up here to play because they have open gym on Sundays and if not then we'll probably just go to Dishman [Matt Dishman Community Centre] or go some other hoop place or the 'Sal', we go to the 'Sal' a lot too.*

INT : Go where?

T : The Salvation Army, we go there a lot.

The male-dominated hierarchies of the park work to marginalize talented young women, because they are seen as lowering the competition level and interfering with the young men's training regimen. In stark contrast to young men such as Adam, Neal, Ricky, and even John, these young women are unable to play in 'informal' spaces such as Woodlawn Park, Peninsula Park and Irving Park because of very systematic codes dictated by the male players.

In the next excerpt Tara describes the differences between the park courts and the school programs:

INT : Can you tell me the differences between like when you play basketball in the parks or in the school programs, do you notice any differences there?

T : Well when you play at the school there is no discrimination against, like your sex, it's just never about if you're a girl or you're a boy, it's just you play, you know. But when you're on your own, people like dudes, when you are at the park, they just look at you like 'oh she's a girl, she's good' or they're just like 'she can't play' or they're like...or they will just kick us off the court. But when we're at school you know, it is everybody's court and not just the boy's court. We put that forward so they share; it's like a lot more friendly.

INT : Where?

T : At the school instead of the park because you don't know the people at the park and they are kind of rude and stuff. But when you're at school you know everybody especially people in basketball, everybody knows you through basketball already because you just know a lot of people, so they are a lot nicer. And it is more organised too.

In contrast to the 'informal' park spaces, the more organised and seemingly exclusive school spaces actually provide more chances for young women to play. Yet, Tara notes that access to space in the school gym is based around perceived capabilities (she says that 'everybody knows you through basketball already') just like the parks.

In a way similar to the park courts, the school gym spaces also operate as sites of male-dominated hierarchies. 'Open gym' is not really 'open' at all because it is dominated by the young African-American men who see it as a vehicle to play 'hard' and 'physical' in order to improve their basketball playing abilities. The only time that young women in North East Portland are regularly allowed to play is during the organised school team seasons or in some community basketball tournaments.

Discussion of Basketball Spaces

In the inner North East neighbourhoods of Portland, basketball is widely considered to be a popular and inclusive activity for young people and a means to a better life. Several high-profile school and club programs exist and Nike has built new courts in many North East Portland parks. These parks and programs, however, are primarily authorized for and used by African-American males who use these spaces in order to improve their basketball playing abilities. While distinctions between 'informal' and 'organised' spaces are often made by researchers of geography, sport, and physical activity, the previous excerpts provide evidence that the definitions and boundaries between these spaces are not fixed, they are actually fluid. Oftentimes it is the 'open' gym or 'informal' park spaces that have the most restricted access, not the seemingly more selective and restrictive 'organised' school spaces. For example, in park basketball there are fewer fouls called, which privileges mostly athletic younger male players who want to 'toughen up' for upcoming league seasons. Picking teams by 'next' is often used as a way to exclude young men and women who are perceived to be physically incapable of helping their teams win games. Park basketball is considered by many young men and even women to be a training ground with different styles and rules. It is also viewed as a means to be recognized in the community and as a way to win high school and club championships, and to obtain college scholarships and maybe even a shot at making it to the professional ranks.

While North East Portland is often positioned as the 'Other' because of higher poverty and crime levels and a large African-American population, within North East Portland basketball spaces there exist numerous 'othering' processes based around physical capabilities and discourses of gender, class and race. Amongst and within the park spaces of North East Portland hierarchies have been constructed whereby highly talented African-American male basketball players are able to freely move between different courts and then 'run' them when they are present. Ricky, Adam and Neal's comments about using various street, park, community centre and school spaces illustrates that they have unrestricted use of their neighbourhoods because they are high-profile male basketball players. They ride bikes late at night throughout the streets of the North East Portland neighbourhoods and they can exclude and play with whomever they choose to in the parks (they usually decide to play with or are picked by other highly-talented young men). Also, these young men do not fear the types of male violence that many of the young women and men with lesser basketball playing capabilities do. In their interviews these young men have also described using a range of basketball spaces that extends much farther than the North East neighbourhoods of Portland and they travel to several prestigious national tournaments across the United States.

Less-talented young men who want to participate in basketball are dissuaded from using city spaces such as parks through threats of violence and a systematic means of choosing teams called 'next'. John, because he doesn't have the same types of basketball abilities as Adam, Neal and Ricky, is only able to play in Woodlawn Park. He can only play in this park because it is very close to his house and he has several friends who play there. John avoids Alberta Park because the 'gangstas' chased him away one time. He wouldn't get picked to play in a game at Irving Park, and he doesn't really know the players at Peninsula Park who tend to be older and not welcoming to outsiders unless they have abilities like Adam, Neal, and Ricky.

Green (1998: 175) argues that sport and leisure spaces are fundamental sites where gender relations and identities are constructed. She posits that sport and physical activity spaces act as social processes which sustain particular gender relations and identities and "work" to produce and perpetuate inequalities and social hierarchies. The young women presented here were excluded or constrained from using the park courts by the young men despite their high levels of athletic ability. While physical capability was often the benchmark for spatial access in the case of the young men, the young women were excluded primarily because of their gender. Even though Sharon and Tara are highly capable basketball players and track sprinters they are still constrained from playing in most of the park spaces. They describe avoiding places such as Peninsula Park (too many 'crazy people' as Sharon says), being physically roughed up during the games, and how they are excluded by 'next' in the parks and school 'open gyms'. Because of this the young women were marginalized and forced to negotiate their spaces in very different ways from the young men - they sometimes sat on their basketballs and watched, they would move to other alternative courts in the park, or they would leave the park to try and play elsewhere. This is similar to the young women in Wilson *et al.'s* (2001) study of an inner city 'drop in' basketball program. In their study the young women would step onto the court to toss a ball around when the young men happened to be at the other end of the court during a full-court game. When the game moved back the young women would return to the bleachers and watch. In relation to the 'male-dominated informal culture of the gym' most of these young women were spectators of all-male basketball games and the young women who did play were severely limited in their ability to play on the gym court (Wilson *et al.,* 2001: 315). Because gender was often the primary criteria for exclusion, young women strategically negotiated urban basketball spaces or found alternative sites in order to play. Through the "complex and dynamic processes of gender-related oppression" (Wilson *et al.*, 2001: 305) operating in the basketball spaces, the young women were provided with different meanings and ways of being.

While the male-dominated hierarchies in the basketball spaces described by the young men and women were (re)produced by the young people (primarily young men) themselves, these hierarchies have also been constructed in relation to prevailing societal discourses emphasizing the individual empowerment of the urban poor (often males) through sport. This was made very evident in comments made by Neal :

> INT : *Why do you play basketball?*
>
> N : *'cause it's fun! It's just, the competition, to see who's better or not. And see what you can do and what you can't do. And like, like basketball's my anti-drug, so, instead of being out in the street, and you know bein' in a gang, I just play basketball, and like, people that mess with me, like gang members and stuff like that, they don't mess with me 'cause they, they know I hoop. And so, it just help me, I don't know, I love it, it's like my life.*

This liberal humanist discourse that is targeted towards the racialised 'urban other' has been taken up by many young people in my research cohort who have seen family members or friends get 'caught up' in criminal activities or in gangs. Here Neal says that basketball is his 'anti-drug' and he has mentioned elsewhere that his cousins are involved in gangs. Ricky's older brother was a talented basketball player who he says 'just chose the wrong path' and spent time in prison. John says that 'lots for me is at stake' and that he wants to make it as a professional athlete because his older brother was a former athlete who went to prison. These young people have a close proximity to gangs and crime and have taken up a dominant 'gangs versus sport' discourse that has been described by Cole (1996). In this dichotomy that has been widely (re)produced by popular media and social institutions, the gang member "whose deviance is expressed through the breach of the work ethic, failed discipline, pathological greed, compulsion, and violence, produces what/who are normal and acceptable: the urban African-American athlete" (Cole, 1996: 370). Through this sport/gang dichotomy, the racialised 'urban other' is further differentiated. Those who participate in sport are "properly disciplined and productive" (Cole, 1996: 367) in the image of the dominant White societies and are constructed in contrast to 'others' who are considered to be deviant, criminal, and 'black'. In this way sports such as basketball are presented to the racialised 'urban other' as a means of being 'straight' and as "the inner city's means for realizing America's utopic promises" (Cole, 1996: 367). At the same time these discourses of 'empowerment through sport' work to discipline and normalize the black or African American 'urban other'.

This 'empowerment through sport' discourse has been taken up by community programs such as Midnight Basketball and high schools. Neal, for example, comments that at Thompson High School 'they show the basketball team so much love' because basketball players are seen as examples of individual academic and athletic achievement. Popular media culture and trans national corporations such as Nike and the National Basketball Association have also (re)produced this discourse of individual empowerment. For instance, in the context of popular media culture Wilson (1997: 181) notes that 'rags to riches' narratives of African American basketball players are widely portrayed as emphasizing "*individual* choice as the key to overcoming difficulties and achieving success". Also, while Nike, Inc. has refurbished the basketball courts in Alberta, Woodlawn, Irving, and Peninsula parks[12] (see Figure 4.3), Cole (1996: 390) notes that Nike and Michael Jordan are both "implicated in the liberal humanist themes of "self made" and "made in America," both are products of trans national capital and its reorganization of America's urban economy and its associated problems". Nike is also considered to be "technologically hip, innovative, and worldly (the American global citizen)" (Cole, 1996: 372) which makes the symbols found on the park courts of North East Portland even more alluring cosmopolitan icons for the young men living in this area.

Fig. 4.3 : Refurbished courts by Nike

Because the 'empowerment through sport' discourse is so prevalent in the inner north-east neighbourhoods of Portland, many of the young men appear compelled to use the wide range of spaces available to them as a means to improve their basketball playing abilities. Through their endeavours and immense investments in basketball they see themselves as trying to achieve a better life. By creating a basketball dream that is tied to individual choice-making and empowerment, sport institutions, governments and 'transnationalized' economies have absolved themselves from dealing with the 'overwhelming poverty and unemployment characterizing the Black underclass in the United States' (Wilson, 1999: 241).

The Issues of Identities

The established societal and popular culture discourses of 'empowerment through sport' are often racialised because they target young black or African-American men such as Adam, Neal, Ricky and John. Pitter and Andrews (1997: 86), for instance, argue that private and public sponsored sport programs such as Midnight Basketball "target underserved youth, who are often referred to as disadvantaged, at-risk, Black, and inner-city youth". As a result, these young men draw from these sport dream discourses in order to assume a black masculine identity.

From Neal:

INT : *What about, you said basketball 'we' like black African Americans, is that right?*
N : *Just as people, yeah black African Americans, yeah.*
INT : *How does that tie to basketball do you think?*
N : *It ties the sports I guess, like that's a game, living here is like not that good or whatever and like then I'll think.*
INT : *What do you mean by that?*
N : *We want better for ourselves and like by using basketball as a tool, football and sport as a tool you try to get out of here, you know like try to get to college, you know try to make something of our lives.*

Young men such as Adam, Neal, John and Ricky all strongly identify themselves in this context as 'Black' or 'African-American' males (even though Adam's mother is from a different ethnic background he takes up a black identity that is tied to his immersion in basketball). Edwards (1983: 32) notes that black men in American society have been 'cut off' from 'mainstream routes of masculinity' such as economic achievement. Because of this success in sport is seen as a means to prove their manhood. Hooks (2004: 94) also posits that because of the high-profile successes of black male athletes, young black men try to "prove their value" and their manhood by participating in sports such as basketball. As a result, basketball spaces are seen by these young men to be sites of "redemption and affirmation" and spaces from which to "gain both visibility and a measure of respect" (Hooks, 2004: 94).

However, by taking up a dominant black masculine sport narrative and identity on offer to them many young men in North East Portland are precluded from conceiving of any other ways of being. Hooks (2004: 22) argues that despite the powerful narrative of black male empowerment through sport, taking up this discourse rarely provides opportunities from which "an alternative masculinity rooted in dignity and selfhood can emerge". For the young men presented here, their African-American or black identities are tied so closely to basketball participation and achievement. Subsequently their comments point to how they are compelled to 'hog the courts' in the parks and gyms and 'play rough' with or exclude other young men and women who are perceived to be interfering with their dreams and identities as sporting black males. For these young men it is nearly impossible for them to conceive of an 'alternative' black masculine identity that might change the way they use the basketball spaces around them and interact with others. This is unfortunate because the young men have constructed and sustained particular gender-based hierarchies that impinge on the ways young women can use spaces and construct their own identities. As Green (1998: 177) argues, "the historical limitation of women's mobility in terms of space, has in some cultural contexts been a means to control and subordination and also a limitation on identity."

While these young men easily take up an essentialist racial

identity that is rooted in sport success and the oppression of 'other' young men and women, the issue of identity for these young men is a complex one. Yet, in the context of their racialised 'othering' by White middle/upper class economic policies and societal discourses it would be hard to imagine these young men not affirming a black male sport identity. Wilson (1999: 232) comments that "Sport, particularly basketball, are sites where young Black males symbolically oppose the dominant White group" and create their particular identities. Because these young men take up such an essentialist black masculine identity that is linked with basketball prowess and Anglo-centric discourses based around notions of individual empowerment, it is nearly impossible for them to imagine not playing in the future. Even though, as Edwards (1983: 32) notes, "the overwhelming majority of young males seeking affluence and stardom through sports are foredoomed to fail".

From Adam and Neal :

INT : Could you both, starting with you Neal, could you see yourself stopping playing?

N : Nope.

INT : Ever?

N : [Shakes his head] 'Cause it's like something that, to me, basketball is like breathing.

INT : It's like what?

N : Breathing. Like If I can't play basketball then, you know, it's like I'm like without air 'cause I love it so much. You know, I'm trying to do whatever it takes for me to play, so, if I couldn't play basketball I don't know what I'd do.

INT : How about for you Adam? Could you see yourself stopping from playing?

A : Naww, I work too hard to stop playing. I work so hard, way too hard to stop playing. And like, to me, I'm like one of the best players in the state, so I work way too hard to, uh, give up playing. So, I couldn't see myself...

Later in the interview:

N : *We work, I mean we do this everyday 365 days a year, so, for us to stop playing that's like 'Why have we been doing this?' you know, that's like our whole, to me that's like...*
A : *Our whole life...*
N : *That's like my life is pointless without basketball, it's not really like that, but it's basically like that 'cause I love it so much it and I wanna do it.*

These comments raise important questions: What happens when the majority of these young men fail to make it to top university and professional basketball programs or become injured? What will be the psychic effects resulting from the loss of this black masculine identity that is so tied to sport achievement? The existential dilemma of choosing an identity becomes more complicated, however, when these young men see black athletic icons such as Michael Jordan who have "been able to pander to the racial insecurities and paranoia of the white majority, primarily because of their ability to shed their black identities in promotional contexts" (Andrews, 1996b: 139). Andrews notes that African-Americans such as Michael Jordan are often compelled to "disavow their blackness" because American culture does not tolerate those who are considered to be "too black" (1996b: 139). Andrews further argues that "Many in the white population are gracious enough to accept, even adulate, African-Americans, but only it they do not explicitly assert their blackness" (1996b: 140). For these young men the quandary of identity construction is to simultaneously take up an essentialist black masculine identity rooted in basketball and also an identity that is not 'too black' (e.g. being the 'threatening racial Other' (Andrews, 1996b: 140) in the gangs and failing to take up the cultural and institutional sport dream discourses offered to them by White society). The consequences of maintaining this requisite balance of identity could be immense for these young men—from their standpoint failure might be seen as precluding them from leaving a life that is, as Neal says, 'not that good'.

NOTES

1. In 2004-2005, 87.5 per cent of Thompson students identified themselves as American-Indian, African-American, Asian-American, or Hispanic-American. 67 per cent of these students considered themselves to be African-American (Portland Public Schools, 2005).
2. From a post-structural perspective it is problematic to use Global Positioning System (GPS) technology to 'surveil' the everyday movements of young people - the use of an invasive surveillance device to monitor individuals' behaviours would be a modern-day version of the panopticon described by Foucault (1979). Through the use of GPS devices young people would be aware that they were being 'observed and on display' under 'the gaze of authority' (Tsang, 2000: 50). Also, Crouch and Matless (1996) posit that geographic encounters cannot be understood in a linear, orderly manner, since different physical activity spaces are psychically and materially experienced by subjects in 'kaleidoscopic' fashion. However, the use of GPS devices might have streamlined the process of developing maps. A drawback might be that these devices are very expensive (approximately 1500 USD each) and would be difficult to distribute and collect on a regular basis.
3. Cole (1995: 347) notes the prevalence of these cultural symbols in the American psyche: "the Nike swoosh, the directive "just do it", and Nike signature advertisements are, indeed, everywhere".
4. The names of all research participants have been changed to conceal their identities.
5. The name of the high school has been changed to provide further anonymity to research participants.
6. 'Crackin'' is used in this context to denote a space which is full of people with lots of activity.
7. Midnight Basketball was originally developed in 1986 and was designed as a means to keep young black males off the inner-city streets, since it was assumed that they would either be committing crimes or be victims of criminal activities during later hours (Pitter & Andrews, 1997). Pitter and Andrews (1997: 89) note that "It is now being suggested that many of today's youth only can be reached and helped by new programs such as Midnight Basketball".
8. I have seen this system of using 'next' or 'down' to pick teams being appropriated by soccer players who play during school lunch periods in a Thompson high school gym. In future analyses I will investigate how this group of soccer players, who are already marginalized (they can only play in an older and smaller gym, in a grass patch behind the school, or even in school basement hallways), use this system to choose players with similar or dissimilar criteria.
9. Between the mid-1980s and mid-1990s Alberta Park was the 'turf' line between the 'Woodlawn Park Bloods' and the 'Kerby Street Crips' gangs and was considered to be a very dangerous park. Recent shootings on successive days

in July 2005 next to Alberta Park show that this can still be an unsafe space (Beaven, 2005).

10. Many of the young women in the study cohort had reservations about going to Peninsula Park. One young Haitian woman called it 'ghetto' and 'people would be just fighting over stupid stuff'.
11. The boys don't 'check' (have concerns or cause a commotion) when they think the young women can play basketball at a high level.
12. Cole (1996: 380) describes how 'Nike's Environmental Action Team (N.E.A.T.) developed a program to collect and recycle old Nike shoes as indoor and outdoor sport courts, playgrounds and tracks: "larger segments of youth are provided with space to participate in sports and physical activity through environmentally sound practices which used shoes, that would otherwise contribute to landfill crisis, are recycled." She considers this program as emblematic of the ways Nike positions itself as a conscientious and compassionate corporation, further contributing to a discourse of individual choice-making and responsible citizenship.

REFERENCES

Aitchison, C. 1999. "New cultural geographies: the spatiality of leisure, gender and sexuality", *Leisure Studies,* 18(1): 19-39.

Aitchison, C. 2001. "Gender and leisure research: The 'codification of knowledge'", *Leisure Studies,* 23: 1-19.

Andrews, D.L. 1996a. "Deconstructing Michael Jordan: Reconstructing postindustrial America", *Sociology of Sport Journal,* 13: 315-318.

Andrews, D.L. 1996b. "The fact(s) of Michael Jordan's blackness: Excavating a floating racial signifier", *Sociology of Sport Journal,* 13: 125-158.

Baudrillard, J. 1988. *America.* London: Verso.

Beaven, S. 2005, August 13. "Community concerns echo downtown", *The Oregonian,* p. C1.

Coakley, J. J. 2001. *Sport in society: Issues and controversies.* New York: McGraw-Hill.

Cole, C.L. 1995. "Celebrity feminism: *Nike style,* post-Fordism, transcendence, and consumer power", *Sociology of Sport Journal,* 12: 347-369.

Cole, C.L. 1996. "American Jordan: P.L.A.Y., consensus, and punishment", *Sociology of Sport Journal,* 13: 366-397.

Crouch, D. 2000. "Places around us: embodied lay geographies in leisure and tourism", *Leisure Studies,* 19: 63-76.

Crouch, D. & Matless, D. (1996). "Refiguring geography: the parish maps of common ground", *Transactions of the Institute of British Geographers New Series,* 21(1): 236-255.

Dawdy, P. 2000. "Alberta rising?" *Willamette Week Online.* [Online]. Available URL: wweek.com.html.leada090600 (accessed 16 September, 2003).

Edwards, H. 1983. "Educating Black Athletes", *The Atlantic Monthly,* 253(2): 32.

Foucault, M. 1973. *The Birth of the Clinic*. London: Tavistock.

Foucault, M. 1979. *Discipline and Punish: The Birth of the Prison*. New York: Vintage.

Green, E. 1998. " 'Women Doing Friendship': an analysis of women's leisure as a site of identity construction, empowerment, and resistance", *Leisure Studies,* 17: 171-185.

Hellison, D.R. 1995. *Teaching Personal and Social Responsibility Through Physical Activity*. Champaign, IL: Human Kinetics.

Hokuwhitu, B. 2003. "Race tactics: The racialised athletic body", *Junctures,* 1: 21-34.

hooks, b. 2004. *We Real Cool: Black Men and Masculinity*. New York: Routledge.

Jamieson, K.M. 2003. "Occupying a middle space: Toward a mestiza sport studies", *Sociology of Sport Journal,* 20(1): 1-16.

Kelley, R. G. B. 1997. "Playing for keeps: Pleasure and profit on the postindustrial playground", in W. Lubiano (ed.) *The House That Race Built: Black Americans, U.S. Terrain,* New York: Pantheon Books.

Martinek, T. J. & Hellison, D. R. 1997. "Fostering resiliency in underserved youth through physical activity", *Quest,* 49(1): 34-49.

Mowl, G. & Turner, J. 1995. "Women, gender, leisure and place; towards a more 'humanistic' geography of women's leisure", *Leisure Studies,* 14(2): 102-116.

Pitter, R. & Andrews, D.L. 1997. "Serving America's underserved youth: Reflections on sport and recreation in an emerging social problems industry", *Quest,* 49: 85-99.

Portland Public Schools. 2005. *School Profiles*. Available URL: http://www.pps.k12.or.us/schools-c/profiles/?id=118#statistics (accessed 26 July, 2005).

Rose, G. 1995. "Place and identity: a sense of place", in D. Massey & P. Jess (eds.) *A Place in the World? Places, Cultures and Globalisation,* Milton Keynes: Open University Press.

Scraton, S. 1994. "The changing world of women and leisure: feminism, 'postfeminism' and leisure", *Leisure Studies,* 13(4): 249-261.

Scraton, S. & Watson, B. 1998. "Gendered cities: Women and public leisure space in the 'postmodern city'", *Leisure Studies,* 17(2): 123-137.

Shogan, D. 2002. "Characterizing constraints of leisure: a Foucaultian analysis of leisure constraints", *Leisure Studies,* 21: 27-38.

Suo, S. 2001, 31 October. "Suburbs suit Portland, Ore., residents of color, minorities snap up homes", *The Oregonian*.

Tsang, T. 2000. "Let me tell you a story: A narrative exploration of identity in high-performance sport", *Sociology of Sport Journal,* 17: 44-59.

United States Census Bureau 2000. *United States Census 2000*. Washington, D.C. Available URL: http://www.census.gov/main/www/cen2000.html (accessed 15 December, 2004).

United States Department of Health & Human Services. 1996. *Physical Activity and Human Health: A Report of the Surgeon General*. Atlanta, GA: U.S. Department of Health and Human Services, Centres for Disease Control, and

Prevention, & National Centre for Chronic Disease Prevention and Health Promotion.

Wilson, B. 1997. "'Good Blacks' and 'Bad Blacks'", *International Review for the Sociology of Sport,* 32(2): 177-189.

Wilson, B. 1999. "'Cool pose' incorporated: The marketing of black masculinity in Canadian NBA coverage", (pp. 232-253) in P. White & K. Young (eds.) *Sport and Gender in Canada,* Ontario: Oxford University Press.

Wilson, B., White, P. & Fisher, K. 2001. "Multiple Identities in a Marginalized Culture: Female Youth in an "Inner-City" Recreation/Drop-In Centre", *Journal of Sport and Social Issues,* 25(3): 301-323.

Chapter 5

CONTESTED SPACE

Children's Use of Space In and Around a South African Squatter Camp

Jill Kruger

Canaansland was the largest of 61 squatter sites in downtown Johannesburg (South Africa). Over a thousand people, including about 142 children aged 0-16 years, lived there in makeshift shacks (*amakhukhu*). A number of parents had sent their children to live in less congested environments with relatives. Adults ran small businesses on site, worked as domestics or manual labourers in the neighbourhood, collected and sold scrap metal, or did piecework. Very few were in salaried employment. The site accommodated about 55 people per 250 square metres whereas this was the official space recommendation for a single family. It seemed impossible that children could access space to play, to study, or to socialise in such a crowded setting.

In 1997, 23 boys and girls aged 10-14 years took part in "Growing up in Cities" (GUIC) - a UNESCO - driven project which enables children to evaluate their environments and to recommend improvements (Chawla, 2002; Swart-Kruger, 2000). The GUIC study was followed by ethnographic fieldwork on site since ethnography yields depth insights through: "... its focus on detail, and its unfailing respect for context in the making of any generalization, as well as the full recognition of persistent ambiguity" (Marcus and Fischer, 1986: 159).

Interaction with the girls and boys from Canaansland in their home settings opened up communications and observations which extended the insights gathered during the GUIC study. It facilitated a reflective process, which drew upon an:

> "archive of implicit recollection and habituated knowledge.... In real life there are no beginnings or ends to particular events; there are antecedents and successors to every moment ... We ... punctuate this in narrative" (Hastrup, 1995: 183-184).

Place and Space as Social Construction

Callaway (1993) proposes that anthropological conceptions of space have three facets. Physical space comprises all natural and human constructions, including homesteads, cities, thoroughfares, or physical barriers such as hedges and trenches. Social space has to do with the ordering of relationships, for instance: kinship structures, socio-economic organisation, labour division, and the domestic lifecycle. Metaphysical space embraces cosmologies or world-views which prescribe and legitimate the way that social and physical space is ordered. The symbolic ordering of metaphysical space might take precedence over, or restrict, the utilitarian use of physical and social space.

The notion of space as socially constructed is now axiomatic in anthropology, and it could be argued that Callaway's categorisation overly compartmentalizes concrete and abstract space. It is not clear where they intersect. For this reason, Michel de Certeau's (1984) clarification of place and space, which hinges on a theory of practice, was preferred when working with children. De Certeau construes place in the same way that Callaway construes "physical space" but De Certeau sees place as transformed into space through human presence. In other words, space is frequented, or "practised place" (De Certeau, 1984: 117) - the pedestrians in a street transform it from a place (as defined by planners) into a space (*cf.* Augé, 1995). Hence places can be thought of as collections of coexisting elements, and space as the ordering and animation of such places through human activity (Augé, 1995) which is underpinned by social structuring, whether this derives from metaphysical, utilitarian or social motivations.

Space which has been socially constructed cannot be neutral and places can become invested with meaning and trigger positive or negative emotion:

> "Feelings about ... people marked as different, may ... be associated with places. Nervousness about walking down a street in a district which has been labelled as dangerous, nauseousness associated with particular smells, excitement ... may be ... sensations engendered...." (Sibley, 1995: 3).

Physical objects influence the nature and use of accessible space, hence also human behaviour, and they may be used intentionally to create a certain ordering of individuals, for instance, place seating at meal times (Ardener, 1993). Whether their placing is incidental or contrived, physical objects always "impress a part of themselves" when they are present (Goffman, 1979: 1).

Hierarchical relationships are inherent in the social construction of space since access to, or exclusion from space, is predicated on shared understandings of status that hinge on interlinked variables such as age, gender and time (Ardener, 1993; James, Jenks and Prout, 1998). The notion of hierarchy always implies inequality rather than mere difference, for differences can only be examined between those who are categorised as equal (Miller, 1993).

As regards age, hierarchy frequently implies power by seniors to exercise domination and control over juniors. In gender and time hierarchies, women may be dominant in one sphere or time, and men in another (Miller, 1993). There is clearly a wide range of dynamics which leads to links in any hierarchy being "open or closed, strong or weak, dynamic or static", and the process of repositioning which is frequently at work, is perceptible in redefined access to space (Miller, 1993: 7). The different status rankings that people hold simultaneously, being junior in certain relationships and senior in others, are perceptible through a number of indicators, including spatial access.

These principles are as true for boys and girls as they are for adults, but children's access to, and use of, space is usually controlled by adults. We also know that children do not experience their physical and social locations in exactly the same way as adults do as they go about their home, school and work commitments. Moreover, through play, children are brought into:

> "intimate contact with their physical environments, often in ways quite different from how adults experience and use environments that are objectively the same" (Stephens, 1994: 6).

Space in Canaansland

> "For children, the neighborhood is more than a physical setting. It defines a social universe. Children, like the elderly, have a particularly heavy investment in the neighborhood environment. Because they are minimally mobile and spend a relatively little time away from the area in which they live, neighborhoods play a special role in children's daily lives" (Berg and Medrich, 1980: 320).

A comparison of Canaansland children's drawings of the area in which they live, with an architectural drawing of Canaansland, shows how the children perceive their immediate living environment. A wealth of data lies behind each of the children's pictures which, though seemingly fragmented, encapsulate kaleidoscopic views of the world as they fashioned it at the time.

Conceptions of the Local Area: Canaansland Boys' Drawings

In Figure 5.1, the architectural, to-scale drawing, reflects congested but precise arrangements of shacks with a main thoroughfare and various small lanes between shacks. Given's drawing, in contrast, features an untidy assembly of dwellings in no particular order. He depicts Canaansland as a small island in the midst of hazardous arterial roads. Their size reflects the enormity of their danger in the eyes of the children who had to contend with heavy traffic, dangerous driving, speeding taxis and vehicles that ignored red traffic lights. Crossing roads was, consequently, very stressful and time-consuming.

Boys confirmed that the railway and shunting yards which are a dominant feature in Given's drawing were exciting exploratory spaces with their tracks, rubble, uncultivated stretches of wild grass and bush and a steep water tower to climb. The station also had an utilitarian function for the many children from Canaansland who used the toilets there, although the station was up a steep embankment, since they had none at Canaansland.

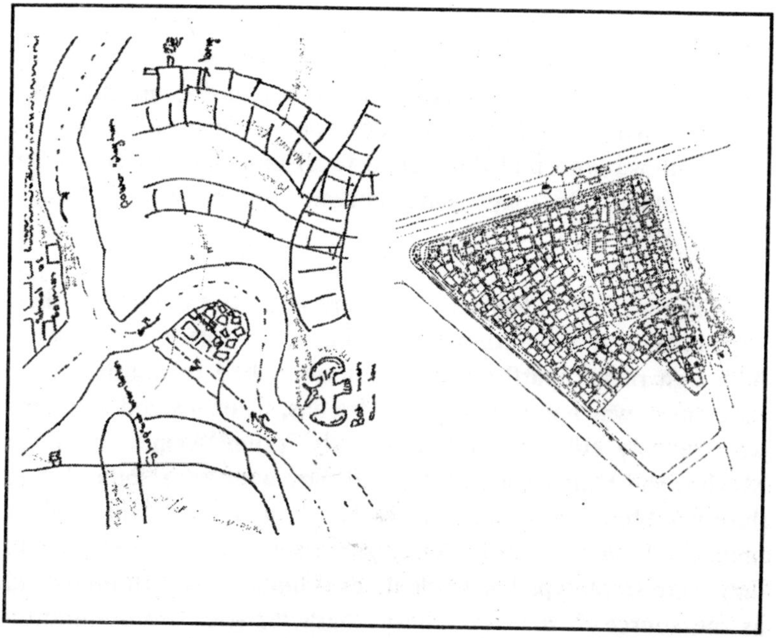

Fig. 5.1 : Canaansland, by Given, boy, 14 years and by Peter Rich, Architect

Given's portrayal of the School of Solomon on the top left side of his map provides vivid evidence of social exclusion by the wider community. He longed to be enrolled at such a school which he considered the epitome of learning with its manicured playing fields, sturdy buildings and pupils with proper satchels and books. He was not enrolled in school at the time and stood fascinated on the pavement, watching the pupils through the fence during their school breaks. One of his favourite pastimes was "...to see the Indians play."

Coercive at core, the exercise of social exclusion is legitimated through institutionalised power where persons are judged, and found wanting, in terms of perceived moral norms (Davies and Morris, 1993). Sibley (1995: 3-4) describes the dynamics at work between dominant and muted segments of society in terms of "landscapes of power" and "geographies of exclusion":

> "Who is felt to belong and not to belong contribute in an important way to the shaping of social space ... often ... hostility to others is articulated as a concern about property values but certain kinds of difference, as they are culturally constructed, trigger anxieties and a wish on the part of those who feel threatened to distance themselves from others. This may ... have economic consequences" (Sibley, 1995: 3).

The residents of Canaansland were socially excluded by their neighbours and the children felt this keenly (Swart-Kruger, 2002). Although racial apartheid had been abolished in South Africa, integration was not effected. As Africans, Canaansland residents were racially, culturally and religiously "other" than their Indian Muslim and Hindu neighbours, and they were also "other" than their neighbours in suburbs to the North, who were white. Their undesirable racial identity was exacerbated by their poverty and they were stereotyped as unclean, as drunkards, as criminals and as the source of disease. These characteristics are commonly attributed to squatters (Sapire, 1991). Their neighbours exercised power as property owners and petitioned local and regional government to have them relocated (see Swart-Kruger, 2000).

The children of Canaansland felt the brunt of the settlement's social exclusion. They were unwelcome in local commercial centres and could not access children's facilities and schools in the neighbourhood. It was hardly likely, therefore, that they would meet local boys and girls in settings conducive to the generation of close friendships, including visits to family homes. Their network of friendships remained insular.

Given's father was in paid employment and he received pocket money regularly. He spent it on snacks and playing arcade games alone or with a friend. The games shop is shown on the left of his drawing; his favourite games were *Pac-Man* and *Thunder Dragon*. Occasionally his father took him browsing at a toyshop in Braamfontein, a nearby suburb, shown in the forefront of his drawing. Sometimes his father bought him toys on such visits. Many boys in their early teens spoke of periodic and unexpected outings with their fathers: to sports venues, to town, to see relatives or to visit some place their fathers thought they would find of interest. They valued these outings and abandoned whatever else they had planned when called by their fathers, even if their fathers were normally abusive and rejecting.

Whereas Given reflected the insularity of Canaansland, his friend Chris divorced himself entirely from his daily living environment when drawing "The area where I live" (see Fig. 5.2). He chose rather, to show a former friend's family home in the suburbs which he fantasised visiting but no longer could. He then focused on a public swimming pool which he and Given sometimes frequented despite white racist hostility, and also drew the local cinema two blocks from Canaansland which he visited alone or with Given or another friend about once a week. As an aside, he wrote a note on his drawing to say that he played soccer behind the shacks in the afternoons. The location for the soccer games was actually four blocks from Canaansland and Chris was the only boy from Canaansland at the time of the study to join youths outside Canaansland in a regular local activity. Interaction was confined to the informal games on a vacant plot of land, however. The other African players were in their late teens and early twenties and learning a trade or seeking work locally.

Here we focus especially on the swimming pool since Chris's image drew strong responses from most of his peers. Despite parental admonitions not to wander about neighbouring suburbs,

Fig. 5.2 : Canaansland drawing, by Chris, boy, 14 years.

some girls, but mostly boys, walked two kilometres to swim at the Brixton public pool which cost sixty cents to enter. Many of them also went to the public pool in *Fietas* which was much closer to home and where there was no entry fee. Nobody knows where the name *Fietas* originated; it referred to the small suburbs of Pageview and Vrededorp which lie between the Brixton and Braamfontein cemeteries. Some of the older children at Canaansland understood *Fietas* to be a derogatory word for the "poor white" residents of these suburbs.[1] *Fietas* was forbidden territory for the children because their parents feared they might encounter white racist hostility there. Children from Canaansland mostly saw the area as an escape from the mundanity of their daily lives and it became invested with danger for them only once their presence as undesirable "others" had been remarked upon.

> "Fietas is a place with parks. There is only one park. At the park we play with swings and swim. I go there with Nokuthula and Zukiswa. I only have friends at Canaansland" (Promise, girl, age 12 years).

> "There are a lot of other kids at the pool and it is never empty so you always have company" (Thembelani, girl, age 10 years).

The stress that impacts on children through social exclusion is keenly felt, the more so because it is personally undeserved, usually unexpected, and violently expressed:

> "... the [adult] person got into the water but couldn't get out. He was drowning... Then one boer [i.e. a white, Afrikaans-speaking man] came running, threw his tee-shirt on the ground ... then got in and got him out... one Boesman [i.e. a Coloured man] came and asked us where do we come from and things like that. And then [later] I went back with small boys. At home I didn't tell anyone where I was going. We got there, took off our clothes, got into the pool. We played in the pool and when we were supposed to go some boers came. They called me and they said "You, tsharo!" [fellow] and they started beating me up. I thought they were playing but they were serious. They came, a whole lot. I thought one of us provoked them and we left. The following day when we came, they did the very same thing to us. Then we realised that we could get hurt and it is not wise if you don't tell them at home where you are going" (Given, boy, age 14 years).

As was the case with Given's and Chris's drawings, the other boys' area drawings did not focus in on Canaansland, but beyond it. Girls' area drawings, on the other hand, showed Canaansland as central and their activities as focused around it. This was not only due to issues of social exclusion but also to common conceptions of childhood held by adults within Canaansland.

Canaansland Adults' Conceptions of Childhood

The daily activities and interactions of children and youth at Canaansland were played out within an ethnic conglomerate of language and custom, which extended into their own homes when children had mixed ethnic parentage. Although Canaansland was ethnically heterogeneous, it was predominantly 'African' and adults were largely in agreement about the behaviour of children and the degree of personal agency which they could be allowed to exercise. There was a strong emphasis on respect for seniority and on unquestioning obedience to reasonable elders. Children were not expected to obey the whims of drunkards or of older children who were mere bullies.

Boys and girls were considered to pass through three developmental stages before reaching adulthood: The earliest stage ended at six or seven years of age when children were believed able to understand the consequences of their actions:

> "Children begin to understand things from about seven years old but you can't trust them with anything much even then. If you catch them playing with matches, you have to give them a hiding because they might burn the shack down" (Amon's aunt. Translated from isiXhosa).

The second stage lasted only a few years until children were ten to twelve years old. They did not yet understand the link between immaturity and a fascination with earthy, natural things, but they became able to carry out instructions, do chores and run errands.

> "Children have to be at least ten before they are allowed to light paraffin stoves, candles, or lamps, but it is even better if they don't do it before they are twelve. Then they slowly begin to take on more responsibilities. Before they are teenagers girls are able to start helping their mothers to cook" (Amon's aunt. Translated from isiXhosa).
>
> "From ten or twelve years old boys and girls really start becoming responsible. You can see the change in their behaviour" (Mother of Bongani and Phumzile. Translated from isiZulu).

The third stage lasted during pre- and early adolescence. Girls and boys were considered able to focus increasingly on formal responsibilities and on the mechanisms of appropriate behaviour in social interactions rather than on the mechanisms of play. Adulthood commenced at about eighteen to twenty-one years. At this stage girls and boys could be employed, were able to carry out tasks responsibly and well, and were thought to be able to train young children competently. These stages are very similar to Western developmental stages of early and middle childhood, adolescence and adulthood.

Adult residents felt that time and place restrictions at Canaansland were important in terms of children's ages and gender. They linked their understandings with tradition and their own upbringing:

> "When I was a child parents ... looked after their daughters carefully, close to home. Even today, I do this myself, although it is difficult here in the city" (Mother of Nokuthula and Zukhiswa. Translated from isiXhosa).

This mother took little note of the comings and goings of her two teenage sons who slept in a lean-to at the side of her shack. She cooked for them and put their food aside in a pot and they ate whenever they wished. Their sisters washed their dishes, did their laundry and cleaned their sleeping place. Neither of the boys attended school and they obtained an irregular income from piece-work in the neighbourhood:

> "If you ask me about the boys, they are now old enough to look after themselves and I do not worry about them. The eldest one [recently 18 yrs] is a young man (insizwa) now and he is able to look after and guide his brother [16 yrs] who is almost a young man too" (Mother of Nokuthula and Zukhiswa. Translated from isiXhosa).

Young boys below ten years of age and girls of all ages knew that they had to remain close to home to run errands and undertake odd jobs for adults. From about twelve years old girls were also expected to help prepare the evening meal. The common response of girls, boys and adults at Canaansland to the question: *Are children important? Why?* was: *Yes. They can be sent by adults.* Some parents said that they also valued having children because of the pleasure that they gave and many said that an adulthood without children would be boring. These insights were secondary to the primary one of having children who could be sent. *Lefura la ngwana ke ho rongwa* (Northern Sotho idiom: The worth of a child being able to be sent - Chairman, Canaansland Residents' Committee) :

> "They can be sent to the shops and for water. When I'm sick they can help or they can run to call other people who can help. When I am dead, they can take care of what is left behind" (Mother of Peter, Belle and Thandi. Translated from isiZulu).
>
> "Mothers want us to stay near home so that adults can send us to run errands. For example, if your parents need something for cooking or anything in the house, they are able to send you while they continue with other chores" (Moses, boy, 9 years. Translated from se Sotho).

On a sunny Friday afternoon, as ten-year-old Bongane sat outside the family shack with a friend, making himself a new *kettie* (Afrikaans: slingshot), his father sent him to buy a single cigarette (*iloose one*). As he handed over the cigarette his father found he was without matches and sent him off again. His friend kept him company. Within half an hour of his return, his mother sent him to buy a bottle of paraffin which he fetched alone. Scarcely ten minutes later, a neighbour sent him to buy a can of coca cola and his friend joined him again. He received three cents as a tip, tied his money in the corner of a cloth which had other similar knots, and hid it in a secret nook in the family shack. Children built up small hoards of spending money in this way. Then he took a message to a friend of his mother's at the other end of Canaansland. By this time it was dusk and the boys went off with the still unfinished *kettie* to sit on the kerb in High Street and complete it before dark.

Children could not refuse to run errands for unrelated adults; only their parents could intervene. Girls and boys would make sure that their younger brothers and sisters remained close to home so that they would be first in line to run errands while they tried to engage with friends out of sight but within earshot of their homes in case their absence was noted. Adolescent boys were sent on more important errands that could not be entrusted to younger boys or to girls. They were proud of this: "By the time that they are teenagers, boys are clever enough to be sent on important errands and to carry important messages" (Amon's aunt).

Teenage boys were, however, not particularly happy if their parents sometimes sent them out into situations which they personally considered unsafe:

> "There is a difference between what they [parents] think is right for boys and for girls, because you as a boy, they would send you to far-away places, anytime, even if it's during the night. But a girl, they won't send a girl. They will say to her, at a certain time she should be back in the house - but you, as a boy, they will say 'you can come in anytime'. They say you are a boy; you can take care of yourself, because you are a gent (ijita). They won't bother with you" (Given, boy, 14 years).

Parents said, however, that they did not keep girls and younger boys close to home only to make use of them. They also feared for their personal safety. Very young children could be hurt in the traffic and children who wandered about the neighbourhood could suffer rape or sexual harassment or be kidnapped and killed for *muti* (isiZulu: medicine).[2] Boys seemed to be treated more leniently than girls if they strayed off site to play with friends.

Conceptions of the Local Area: Canaansland Girls' drawings

Figures 5.3 and 5.4 illustrate the centrality of family shacks for girls in Canaansland when they frame their neighbourhood. Zukiswa (Fig. 5.3) shows streets as links to specific places that are important interactional spaces in her daily life. These are utilitarian in the first instance but she has learnt to experience some of them as spaces of exploration and personal enjoyment:

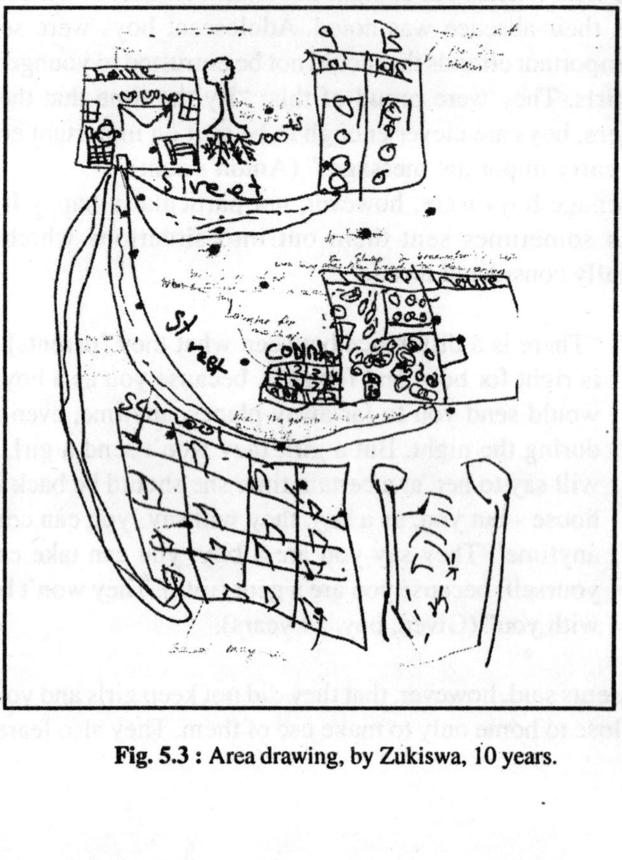

Fig. 5.3 : Area drawing, by Zukiswa, 10 years.

"I like my home. It has a place to sleep and to cook and to eat. There is a tree outside where I play alone or with friends and I sit there if it is hot. These toilets are at the station. There are also toilets at the bus stop near Bree street. I use these toilets mostly. This is my favourite shop. I go there twice a day when I am not at school. We buy eggs, rice, bread, tomatoes and onions here. When you get into this shop you are given a card then you go to the counter with the same number as the card. I enjoy my school. I enjoy learning. They also give us food there. There are netball posts where we can play netball" (Zukiswa, 10 years).

"Myself, money, sometimes I get it if I am sent to the shop. Maybe I know where the thing that I have been asked to go and buy is cheaper. I'll go and buy it there and then keep the five bob [small change] for myself; then I'll go and buy cheese Simbas [cheese-flavoured crisps] and peanuts" (Zukiswa, 10 years).

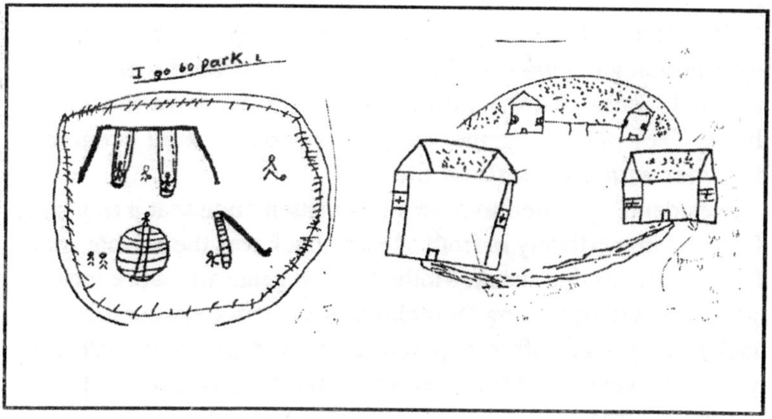

Fig. 5.4: Area drawing, by Phindi, 10 years.

Phindi (Fig. 5.4) shows the homes of other friends in the group: Zukiswa, Nonhlanhla and a girl called Promise. Zukiswa's elder sister, Nokuthula, was often part of this friendship group. Like Zukiswa, Phindi does not depict the less significant shacks roundabout. Phindi shows Canaansland as a bounded area and explained that this was where all her daily activities were located. Phindi was not enrolled in school. She felt that she had a good relationship with her mother who allowed her a great deal of freedom in their shack. However, she considered many Canaansland residents to be dangerous people because she had seen some of them drink too much and stab each other. This saddened her. She longed for a home with electricity and running water which would relieve her of tedious chores. The park in Brixton played a significant part in Phindi's life. She yearned to go there daily and would visit it with her friends if they did so on weekends. Sometimes she would go alone if her mother was out for the day. Phindi found the Brixton park to be the only place which she could personalize as a space for socialization and exploration. Its significance in her drawing is equivalent for her, to the whole of Canaansland. Among the girls, she was the only one to depict a place outside of Canaansland when asked to draw "The area where I live."

Phindi and the other girls in the friendship group commonly went shopping together for their families and shared reading time in their homes (shacks) with books borrowed from the school library. They played together and, being popular, were in demand for games with other girls on site.

Residents appeared to have an unwritten code that a tiny strip of ground immediately in front of each shack was the private space of the owners of that shack. Mothers could generally work indoors and feel secure in leaving their children immediately outside their shacks. They were often exposed to the sun or other inclement weather, however, and this was why Zukiswa particularly valued the family's tree.

Children engaged in a variety of games that enabled them to remain within small spaces and these games often entailed gender separation. Zukiswa, Phindi and the other girls detailed the games that they played outside their shacks or on the pavement within earshot of their mothers. Children who played on the pavement were invariably subjected to verbal or physical abuse by passersby who wanted to belittle them as squatters. Their stones or toys might be scattered and they could be shoved into the street despite the traffic.

Girls played *diketo* almost everyday. This game required dexterity to toss and catch a stone while a pile of others was simultaneously moved about within a small circle. Card games were popular and, if they played outdoors, women they knew and liked, might join in. Boys also played card games but girls preferred to exclude them from their games on grounds of cheating. Boys alone played games with marbles and roved about Canaansland with homemade slingshots. Girls and boys played "house-house" or "school" together in empty shacks which awaited new tenants or in specific children's homes where they could drape a blanket over furniture to create a closed space:

> "Near the table in the kitchen, is our space where we play house-house, where one child pretends to be mum and another one will be a child Outside behind the house, I keep my valuables that I collect for playing house-house; sometimes I put them inside behind the door before we go to bed. Sometimes my mother throws them away by mistake" (Nonhlanhla, girl, 11 years).

Contested Space

The only public space in Canaansland was the central thoroughfare (see architectural drawing, Fig. 5.1) and this was claimed and monopolized by adults. None of it was demarcated for use by children. All children's use of public space at Canaansland was contested space, yet girls of all ages and boys, until the age of 10 or 12 years old, were expected to lead relatively bounded lives there.

Sometimes late in the afternoon or on weekends, when most girls and boys were on site, pockets of playing children would break into noisy, active games which drew in other children and spilled across Canaansland. Games such as chase, *vrot* (Afrikaans: rotten), and hide-and-seek all led inevitably to wild activity and excited screaming. Girls and boys, assured of safety in numbers, would tear up and down the main thoroughfare usually reserved for adult male pursuits such as *morabaraba* (a board game widely played in Africa and Asia), discussions and drinking. They would dash between shacks threatening to bring down shaky structures. Threats and scolding could not stem their tide and the game would end only when they were tired or if a child was unlucky enough to be caught and beaten by an angry elder. Despite the fact that children used the thoroughfare in defiance of adults at least twice a week, no attempt was ever made to re-assign this space so that children could have fair use of it. Adults took little notice of the few toddlers who played there in isolated spots in the dirt periodically during the day, provided that they were not noisy. If they ventured too close to the adult men they might be scolded and run off in fright; they might be picked up and returned to their mothers; or they could be slapped by a drunk, and run home crying.

Boys and girls tried to create a children's play zone in a seldom-used side street (High Street). It was ideal for skipping, hop-scotch and *umgusha* (a girls' rope-jumping game) and boys of all ages kicked a ball about there in the early evenings. Even this space was contested. Although the local council agreed, when requested by the children, to approach the traffic department to close the street off, this never happened while the children lived at Canaansland. Consequently the character of High Street remained ambiguous. It was a tarred road shaded by plane trees, empty of traffic and welcoming to children with no assigned space to call their own, yet officially allocated to occasional drivers who were antagonistic to "squatters" and who had the power to disrupt lives by deliberately inflicting injuries in hit and run encounters.

Extending Boundaries

Girls and boys visited the park and swimming pool in *Fietas* and the swimming pool in Brixton despite periodic problems they encountered there. Those who were enrolled at the school for street children also spent a considerable amount of time at a grassy park in Brixton where there were swings and a roundabout since they could reach this by a short detour on their way home from school. They told their parents that after-school activities had delayed their return home. The lack of telephones and motor cars at Canaansland was a boon to children with glib excuses such as these since it became a major task for parents to check the validity of their excuses. The vast sweep of the park at Brixton provided Zukiswa with a platform on which to role-play fantasies developed from real life scenarios with three friends and her sister: "When we got to the park we would play a story, some will play a Mother character, a sister character, and so on. Yes, that's where I liked to be" (Zukiswa, 10 years).

The girls also played out fantasies by choosing stones that reflected various people's characters. Then they enacted stories around their characters. They sometimes became so engrossed in their stories that they arrived home very late, to a scolding:

> "...if you go [from home], they [parents] will tell you when to come back but for the boy they don't mind what time he will come back. If he comes at twelve, nine or eight, they don't bother him, because he's a boy and I'm a girl, and I know what my times are supposed to be" (Lindelani, girl, 12 years).

Despite these restrictions, girls occasionally broke bounds more spectacularly and explored unknown suburbs regardless of the beating they expected when they returned home. Girls reminisce about one such excursion with keen enjoyment; Nonhlanhla (11 years) described it:

> "...one time ...we went and we met some brothers (guys) and they asked us where we were going and we said we are going to school and they said "So late?!" but they didn't believe us because we didn't have tickets. We got off at the station in Braamfontein, then when we were there we saw the glue boys [streetkids] and they called Nokuthula: "Come here" [crooking a finger]. Nokuthula went and they said "Where's the money?" Nokuthula said: "We don't

> have". And ... Zukiswa said: "We don't have." And they chased us and chased us and they caught Nokuthula and they tripped her, making her fall, and they scared her with a knife, as if they would stab her [background voice: "Hau! You are lying!"]".

Zukiswa and Nonhlanhla shared a different narrative, on an occasion when they had gone with other girls to beg for money in the wealthy suburb of Rosebank in Johannesburg:

> "We slept at the bus stop... we met some grown-ups [raucous laughter from girls] and they were inviting us to their house, to go and cook for them and clean for them. When we have finished, we'd give them some vetkoekies [*lit.* fat cakes: referring to female sexual organs]. Them, they will give us Russians [*i.e.* Russian sausages: referring to male penises] - and they will put on condoms" (Nonhlanhla, girl, 11 years).
>
> "Myself I thought those things were balloons. I asked and they said "Hey you! These things are for grown-ups. They are condoms." That's how they told me. And they told us we will get some money. From there I decided I will take a tin and go and get some money [from begging]" (Zukiswa, girl, 10 years).

It was these dangers from men that parents wanted to protect girls from, among others. When they had to send their daughters some distance from home they would arrange adult supervision for them, so they were very upset and angry when daughters broke boundaries and put themselves in personal danger by going off alone.

> "Busisiwe and me, the time when were learning at Y, we used not to go alone — my mother would accompany us, maybe in the morning and then even go to Y to bring us back in the afternoon. And then we would pass boys who were smoking glue.[3] She would sometimes accompany us to and from a taxi if she could not go all the way with us.

> And then we would be saved from the glue boys who sometimes would chase us and want to take our money or anything we had, from us — because there is no way to come into contact with them when we are not alone" (Phumzile, girl, 12 years).

Although girls and boys periodically went off to places that they knew were forbidden, boys tended, as they passed through their teens, to become more blatant about going wherever they chose and not offering anyone explanations as to their whereabouts.

Conclusion

I have discussed some of the children's issues around private space within their family shacks elsewhere (Swart-Kruger, 2001). This discussion locates children's perceptions and use of space largely beyond the interior of the family shack and takes cognizance of the ways in which girls and boys accepted, negotiated, manipulated or defied adult understandings and directions concerning the social construction of space.

Although Canaansland was unfenced, children's drawings show that it had strong boundaries in the form of hazardous arterial roads which shut children in from the outside world. Social exclusion enforced the insularity of their social interactions and "earshot" defined boundaries for girls and younger boys on site, confining them largely to small pockets of space within Canaansland. Nevertheless, their exploratory urge drove them to push past all boundaries and venture into adjacent suburbs on a fairly regular basis and to venture even beyond those boundaries periodically and place themselves at real personal risk.

De Certeau's (1984) concept of *Zwischenraum* (space that is betwixt and between) is particularly apposite when we consider Canaansland. This no-man's-land with an avenue of plane trees, once a police parking lot, was relatively inconspicuous. It appeared, to homeless people, to be an ideal place to put down roots. As *Zwischenraum,* however, it caught the attention of settled people in neighbouring suburbs as well as developers, once it was seen to be inhabited by squatters who were "unacceptable others". Eventually all people in Canaansland were relocated and low-income housing was developed where they had once lived. *Zwischenraum*, typically, is targeted by developers who want to convert non-places into established place (De Certeau, 1984).

Canaansland was not *Zwischenraum* for the adults or children who lived there. Adults believed that they had converted unused land into established place but Canaansland offered only contested space for children. The nature of their reception into this space was never constant: whether they experienced hostility or benevolence depended very much on individual adult whim.

The children saw the High Street, on the other hand as *Zwischenraum.* They wanted to take possession and transform this planned place that had lost its original utility into their personalised recreational space. However, they were thwarted in this through a lack of personal power to alter local planning regulations. The street remained a place of ambiguity and sudden danger, for children and adults alike.

NOTES

1. In the 19th Century low-income housing was provided for "poor whites" in *Vrededorp* (Afrikaans, meaning town of satisfaction) and Pageview. These suburbs gradually became multicultural but people of colour were forcibly relocated in the 1950s when the area was redeveloped for settlement by low-income white residents (see Carrim 1990).
2. Children's body parts enable witchdoctors (isiZulu: *abathakathi*; N Sotho: *moloi*; isiXhosa: *igqirha*; seTswana: *ngaka*) to make strong medicines and spells because children have drawn but little on their personal luck which is, therefore, still strong.
3. Inhaling glue fumes from plastic bottles.

REFERENCES

Ardener, S. 1993. "Ground rules and social maps for women: an introduction," (pp.1-30) in S. Ardener (ed.), *Women and space: ground rules and social maps.* Oxford: Berg.

Augé, M. 1995. *Non-places: introduction to an anthropology of supermodernity*, translated by John Howe. London: Verso.

Berg, M. and Medrich, E.A. 1980. "Children in Four Neighborhoods: The Physical

Environment and its Effect on Play and Play Patterns", *Environment and Behavior,* 12(3): 320-348.

Callaway, H. 1993. "Spatial domains and women's mobility in Yorubaland, Nigeria", (pp.165-182) in S. Ardener (ed.), *Women and space: ground rules and social maps.* Oxford: Berg.

Carrim, N. 1990. *FIETAS: A Social History of Pageview; 1948-1988.* Johannesburg: Save Pageview Association.

Chawla, L. (ed.) 2002. *Growing up in an urbanising world.* London: Earthscan and UNESCO.

Davies, T.W. and Morris, A. 1993. A Comparative Question of Stigma. *Social Work and Social Sciences Review,* 1(2): 109-122.

De Certeau, M. 1984. *The practice of everyday life.* Berkeley: University of California Press.

Goffman, E. 1979. *Gender advertisements.* London: Macmillan.

Hastrup, K. 1995. "The inarticulate mind: the place of awareness in social action", (pp. 181-197) in A.P. Cohen and N. Rapport (eds.): *Questions of Consciousness.* London: Routledge.

James, A., Jenks, C. and Prout, A. 1998. *Theorizing Childhood.* Cambridge: Polity Press.

Marcus, G.E. and Fisher, M.J. 1986. *Anthropology as cultural critique: an experimental moment in the human sciences.* Chicago: Chicago University Press.

Miller, B.D. 1993. "The anthropology of sex and gender hierarchies", (pp.3-31) in B.D. Miller (ed.), *Sex and gender hierarchies.* Cambridge: Cambridge University Press.

Sapire, H. 1991. *The reaction of established White, African, Indian and 'Coloured' communities to informal settlements.* (Research Report no.19) Johannesburg: Urban Foundation, Informal Settlements Monitoring Project.

Sibley, D. 1995. *Geographies of exclusion: society and difference in the West.* London: Routledge.

Stephens, S. 1994. Children and environment: local worlds and global connections. *Childhood,* 2 (1/2): 1-21.

Swart-Kruger, J. 2000. *Growing up in Canaansland: Children's Recommendations on Improving a Squatter Camp Environment.* Pretoria: HSRC & UNESCO. www.hsrc.org.za

Swart-Kruger, J. 2001. *Isikathi Sokulala*: How boys and girls in a South African squatter camp experience bedtime. *International Journal of Anthropology* (Special issue on children and childhood in anthropology), 16(2-3): 99-111.

Swart-Kruger, J. 2002. "Children in a South African Squatter Camp Gain and Lose a Voice", (pp.111-133) in L. Chawla, (ed.) *Growing up in an Urbanising World.* London: Earthscan and UNESCO.

Chapter 6

AGE AND CONTROL

The Nation, the Neighbourhood and the Home

Nita Kumar

In this preliminary anthropological study of the structure and imagining of space in the lives of children, I go through the following stages of discussion. First, I consider how the construction of an environment is used to discipline the body and the self. The child, who finds herself subject to a certain space in the home, is being taught to be a certain kind of being. This is the politics of age domination. The child can only manoeuvre within it, and it is only as adult that she can manipulate or change it, and then in turn make it an instrument of teaching for her children. Given this positioning *vis-à-vis* space, I feel confident in talking of a specific experience called 'childhood', and in regarding it as a specific aspect of education that can be retrieved.

Two, I discuss the location of the child in the nation, and how an initial difference appears in the imagining of the nation between the schooled and unschooled child. But the difference produced by schooling is offset by the learning that is in common, for instance from the media; and partly offset by the sheer poverty of schooling across the whole range of schools. This pedagogic poverty equals the failure of the nation state. The nation is far more interestingly imagined in films and songs.

Three, other spaces that produce a common experience for all children are the neighbourhood and the city. My city is the provincial city, not the metropolitan one, characterized par excellence by garbage. There are commonalities in the technology of space, as well as in metaphor and imagery, that cut across class.

The neighbourhood, at least of the provincial city, might be one of the most powerful sites for the learning of a common habitus.

Four, given all these commonalities, and turning back to the home, I find that the most crucial lines of division within a class called children are between girls and boys: girls privilege the home, boys the neighbourhood; girls know they have more than one home, boys that they will forever wander. Their personas, and their lives, as male and female, are shaped by the spaces they find themselves in, the spaces as they read and live them, and the spaces they anticipate.

Age and Control : the Fact of Childhood

In seeking to talk about children, I am proposing a belief in some form of 'age culture', and that 'children' or 'childhood' are somehow a reality. As a discursive proposition, I have argued elsewhere, this reality needs questioning for South Asia (Kumar, 2000). However, I will postulate here, as an anthropologist, that a demonstrable 'children's culture' and 'world of the child' exists, in that there is observable action performed by children, games and activities, participation in social life and indeed, lifestyles (for Germany, see Fuchs, 1994). Whether these are accepted and sanctioned by adults is relevant, but not central to the proposition.

What is central is the creation of much of the child's world by adults, or the politics of age. Space, in the home, the neighbourhood, the city, and the nation, is constructed and provided by adults for children. Even when involuntarily done, this involves control and domination of the child, and the effort at disciplining. Benignly speaking, we may call it 'teaching'. The uses of space by adults are all aspects of a teaching strategy by which effective socialization is carried on, and the body is adapted to the constructed environment.

The Case of Zeenat, a Weaver's Daughter

In class three of the Madrasa Hamidia Rizvia for girls, I watched how the teaching consisted of writing answers on the blackboard for children to copy - some explanations having been thrown in for my benefit - and then the teacher going into a kind of doze while children wrote at their own speed, or not at all. Those at the back played, gossiped, and exchanged objects of value. Those in front must have been the better, or best, students, because their entertainment consisted of jerking their pens to form pools of ink, scrambling to share the ink, asking each other for ink, discussing the shortage of ink, and then with a pout bending down to write.

Zeenat was right under my nose so I picked on her. Or, rather the teacher did first, asking for her book and her copy to show me. She would not have chosen a bad student, so the choice of Zeenat was partly because of her location and partly her scholarship. She had indeed a fine hand and also an aesthetic sense, writing all the questions in green and the answers in blue. She looked very serious and would not respond when I tried to make her smile or laugh. Through her teacher I got her to invite me home.

Her house was just across the road, an excellent arrangement by all considerations. She seemed so 'good' that I took for granted that she was from a 'good' family, by which I meant without reflection, a comfortable, middle-class, upwardly mobile family in a clean, spacious set of rooms that I could look at approvingly. We climbed up four flights of steep stairs of a house with an opening in the middle like a well and rooms all around. It was the all-too familiar design of all old homes in Banaras, seemingly more of Muslim homes but that only because more Muslim homes were old-fashioned. We climbed up and up and on each floor women and children gathered and stared at us without reservation (I mean, stared at me).

The room she took me to was equally familiar. There was a cooking space on the landing outside. The room had no furniture and was barely seven or eight feet wide on each side. The father sat on his haunches against a wall. The mother sat against another. A mat was spread against the third wall. The fourth wall was taken up by the door and a window through which peeped family members constantly. In other words, it was cramped and poor.

They had five children, Zeenat being the middle one. The oldest boy, Wasim Riaz, and the girl, Shahina Parveen, had both dropped out of school early because they did not have the inclination for studies. They were bright in other things? I prompted. Neither mother nor father responded, although we had excellent rapport otherwise. Wasim wove and Shahina did housework. Zeenat and her immediately younger brother Mukhtar Naim went to school, she in class 3 in Hamidia Rizvia, and he in class 3 in Islamiya. Rizvia is a Barelwi madrasa and Islamiya a Deobandi one, and regardless of the importance of sectarian differences as presented by maulanas, this family like many, did not seem to regard them as significant.

They were very articulate on the question of some children studying and some not. By and large school education is seen as useless and a burden. While the child studies she can do nothing else, such as sewing or cooking. Because of the burden of studies many families keep tutors also. Anyway, some children 'can' study - they have the *zehen* or inclination - and some 'cannot'. Zeenat and Mukhtar can, and Wasim and Shahina cannot. However, all four are equally accepted. The youngest, I was told, a toddler of two, Shagufta Yasmin, was going to be the best scholar of all—she was already so interested in books. When I asked about the whole place of studies in their lives, Waheeda Bibi explained candidly that nowadays when a marriage was being fixed, the groom's side asked, how far has she studied? Did she complete the Quran? Indeed, in order to make Shahina properly marriageable, she was teaching her at home. The two others had completed their Quran courses.

Quran courses were held in the house of her *bari Amma*, or senior mother, grandmother, or aunt, with at least twenty girls gathered under a maulani. Classes took two hours a day, from 2.30 to 4.30 or 5. There were no charges, but there was a gift given on graduation, and sweets at other occasions, and when the student would get married, a set of clothes. The *ustad* was to be honoured next only to the parents.

Zeenat was sitting close to her mother and looking more serious by the minute, so I tried to speak to her at which she seemed to draw into herself all the more. She must have been so hungry, having eaten whatever she did before 7 a.m. and then carried a mere tiffin to school. But there was no sign of lunch; indeed there were chunks of raw meat showing no input of preparation. Perhaps that was for dinner. So, lunch? I asked carefully. Oh, it's early yet, I was told. We'll eat at 1.30 or 2. They served me Crackjack biscuits and tea. I hurriedly gave the two school children a biscuit each, under protest. The little baby wanted tea and got some.

I turned to Zeenat because I was her guest after all and there because of her. What were her games, her toys? She didn't budge- too hungry? - but her brother brought down a lovely doll's chair and bed, and then a doll. Then he brought down a suitcase full of little clothes stitched by the children, every bit as good as those sold world-wide by companies like Mattel. The baby played with everything as I exclaimed over it. I suddenly realized that Zeenat was too old; by a rough calculation, the youngest was two, the next 10, Zeenat 12, the next children 18 and maybe 19. None looked of their age. All were small built, thin, and tired looking, except the mother who was plump and glowing. Wahida Bibi was educated, since she had spoken of teaching her daughter. But she had not gone to school or madrasa, rather learnt at home from female family members, as they had learnt before them.

She stitched everything at home: her clothes, the children's clothes, the linens, everything. Only the gent's clothes had to be bought. How had she learnt? How would her daughters? She had learnt just by doing. That was how everyone ever learnt. Zeenat was too busy with her studies to sew, but when the madrasa was closed she did whatever she wanted. By such doing, one learnt. Her sister had lots of time once she left school and was an expert seamstress.

The room in which the family lived presented a huge problem. What was the little Zeenat learning? In school, watching her, I had thought that she was destined to study far and was perhaps nurturing ambitions inside her serious little brain and that she would break her family tradition in some way. After all, I had only just met several women teachers, such as Safia Sultana, who had completed her *maulvi, alim,* and *fazil* degrees while living in a hostel in Maunath Bhanjan, sent there by her family in north Banaras, and who had been teaching since, even with two sons one and two years old.

But upon seeing Zeenat's family, I realized that the little girl's seriousness at school was destined to pass like a happy dream. Whatever she ever did, she would do very well. What she did now was schoolwork; what she would do later was housework and child rearing. All the time she was learning as conscientiously at home as in school. In school, she was learning the reading and writing of three languages, and some other subjects. At home, she was learning cooking, sewing, keeping baby, serving, tidying up, running for errands, all the details of how a home works and what in their 'culture' and 'community' a homemaker should do. She had finished with the Quran and although that is learnt only by rote, there was other intangible teaching that goes along with it that she had imbibed.

Zeenat was learning how to deploy herself in the one room her family lives in, to share the bathroom on the landing with the extended family, to cook with her mother, and to cook gradually by herself in the space outside their door, to clean up each time and leave the space unmarked, and otherwise practice a total non-specialization of domestic spaces. The room and its outside were swept morning and evening or when littered, whichever came first. All materials were put away after use. No space was called after its function, as the place 'for' some particular activity, including sleeping. The room became, in turn, a bedroom, a dining room, a sitting room, a den, a study, a workplace, a dining room again, and again a bedroom. Zeenat was mastering the discourse perfectly without a word being spoken on the subject. She was learning further where and how to spread out the rugs for those who would work or sleep, and those who visit. She was learning how every niche and cranny of the room has its uses.

Zeenat is, as a child, and will grow up into an adult as, a being perfectly comfortable and 'natural'. When she is transported into another, foreign space is when she might, miraculously, become awkward and unnatural, as described by Abu-Lughod of the Bedouin family in a glamorous upper class home full of artifacts (Abu-Lughod, 1993: 120). But she will also go to another space that will be interpreted as 'natural', a place that will become 'home' to the extent of her replacing this home with that one. Her married home will be a home configured initially by others, such as her mother-in-law, and then gradually (and dialectically, we hope) both be constructed by her as it is constructing her too. As I note about a different household and its mistress elsewhere (Shree, 2000: 183-86), it is the very design of the house that can impose a discipline on the daughter-in-law. The *mai* house has a central courtyard with verandahs and rooms on three sides, which ensures that the mistress of the house is in confinement because she can always be observed and criticized by the bed-ridden mother-in-law, sitting strategically on one side of the courtyard. If *mai*, our protagonist in this story, had initially had a closed kitchen with modern gadgets and a counter—all of which she got, but too late—she would not only not have had the bent spine she got, but would have had the privacy to order her activities her own way. But there are other, more mixed, results of the house design. In the absence of a master bedroom and the companionate marriage that implies, *mai* has the freedom to spend time with her children, in their room, hers and theirs, and build an inter-generational solidarity. Such a solidarity is the stuff of the socio-cultural training which constitutes 'tradition' in families, equalling a lack of freedoms in a certain perspective but other freedoms in a different perspective. For the daughter in this story, Sunaina, and maybe for Zeenat in the future in ours, there comes to be an identification of the house with childhood, with life itself, with something that haunts and keeps coming back. The house bears the children, it binds them, it makes them grow, it suffocates them. They dream of escaping from it. When they escape, they dream of coming back to it. Homes, even more than parents, control lives, and become for the children the locus of childhood and the past.

Much of the culture that anthropologists locate, including lifestyles, values such as of interpersonal relations, personal identity, honour, and shame, are all aspects of the functioning of the environment and its bio-politics. We have insufficient discussion in Anthropology yet of how spaces create and control lives, even as Henrietta Moore pointed out in her interesting study of the Marakwet of Kenya (1986), "It is now axiomatic that spatial relations represent and reproduce social relations, and it is the view that relations of likeness exist between social distinctions and spatial boundaries that links the study of gender to the study of space" (xi). The thrust of the research on space, or rather, buildings, as a political technology for the exercise of power by organizing and canalizing bodies in space derives its inspiration from Foucault (1975, 1984; Rabinow, 1984). This is taken forth to the case of colonies like India to show how architecture, city planning, and aesthetic forms were used for culturally and epistemologically oriented colonial domination (Rabinow, 1982; Oldenburg, 1984). When brought forward to national periods, the exercise of power by national states is seen in the continuation of the same techniques of architecture and planning of the colonial periods, leading to even more elaborate control of the domains of daily life (Holston, 1989). I am proposing a discussion of a more subtle, everyday politics: the colonization of children by adults that also translates into an extraordinarily strong socialization, resulting in inter-generational continuity in values. The question that remains to be asked alongside is about how children reappropriate space, and what their tactics are, overt and clandestine, to escape the nets of disciplining. How do they themselves participate in their socialization? Because... "the pedestrian's walking, like the flaneur of Walter Benjamin…is the spatial acting out of place, creating and representing public space rather than being subjected to it" (Low, 1999: 113-14).

Useful for my approach is also the work of Bourdieu (1977) and the concept of *habitus*:

> "In his examples, the Kabyle house becomes the setting in which body space and cosmic space are integrated through metaphor and symbolic homologous structures. Through the experience of living in the spatial symbolism of the

home, social structure becomes embodied and naturalized in everyday practice… Bourdieu also examines resistances and the effect of feedback on the social system. Since the concept of habitus spatially links social structure to the human body and bodily practices, the possibility of resistance to these practices becomes more apparent (Giroux, 1983)" (Low, 1999: 114).

Education : Flaws in Teaching the Nation

Because, specifically, of the colonial history of India in the last two centuries and the ways that modernity was defined and used, schooling came to play the part of the archetypical signifier of the nation state, progress, and upward mobility. The space of the school came to be a fetishised one. It became a ritual space where a wonder was supposedly enacted. Because it was, and continues to be, closed and hollow, no one may look inside and almost no one has a precise idea of what is enacted within, especially not the parents. But a magic was/is supposed to occur, indeed, the biggest magic of modernity: education can apparently change a person from one kind of being to another.

One would expect, then, that the child's understanding of space also varies according to whether the child is schooled or unschooled. This interesting proposition of the school as marking the social division in India between two parts of the population can be investigated with regard to the nation.

The nation for the schoolchild is visualized as a geographical space: a map of a certain space that can be brought to life with images of mountains and rivers, plains and countrysides. This map and this collage of images could be encountered routinely by the child in the school, and in a spectrum running from strong to weak to non-existent, its presence gets diluted from the most modern (a missionary school) to the least modern school (a municipal school). The nation, as I have discussed elsewhere (Kumar, 2001, 2002) does not exist as a physical, material concept for the unschooled child.

Further, a child going to school can answer questions about the nation, that is, he can *imagine* the nation. An ironic characteristic of schooling in India is that the more powerful or effective the schooling is, the more it has to ignore the local. Schooling does not so much battle with the local as render it invisible, which is of course one of the strongest symbolic weapons to wield. 'Progress' and 'modernity' are measured with regard to the distance from the local. This distance could carry one well beyond the nation into the terrains beyond (called the global by us, but not my informants), and this is exactly what happens (Kumar, 2001).

Let us look very briefly at the play of national and spatial symbols in three schools: St. Johns, Gopi Radha, and Jamia Islamiya. St. Johns has 'foreign' pictures of romantic landscapes, from the same repertoire as its others of cuddly animals, blonde infants, and bouquets of flowers. It also has a large concrete map of India on the wall, discoloured by being outside where the assemblies are held, and a flag of India that is unfurled on ritual occasions. The 'houses' students are divided up into for competitive sports and other activities are named after national heroes: Gandhi, Nehru and Tagore. Elsewhere, in Loreto, they are Carmel, Fatima, and Lourdes, and in the Ramakrishna Mission School there is Vivekanand and Netaji Bose. When questioned, the students had no idea what role 'houses' were supposed to play in their lives, and the map of India was consensually relegated to the unwanted margins, but both do exist.

In Gopi Radha and provincial schools which are either Hindi medium or English medium only 'in name' (not in fact), there are pictures of Krishna in various contexts, the choice falling, ironically, more on Krishna as adult than as child. There are pictures of Saraswati, the patron of learning and the arts, blessing without discrimination. There are idols closer in history, such as Gandhi, Nehru, and Tagore, always straight-faced and forbidding. There are landscapes of mountains and fountains, the idealized scenery never typically seen by the children. The principle seems to be to maximize the distance from the mundane and the real, and immerse them in an adult world of the desirable. But 'immerse' is hardly the right term because the effort is technologically so flawed. The posters and pictures are not at children's eye levels but at adult height. Anything remotely prone to damage, such as of glass, clay, or paper, is safely removed from within touching distance. There are no colours, patterns, or images in tune with children's mental worlds, and certainly nothing produced by children themselves.

Jamia Islamiya, like all madrasas and Sanskrit schools, eschews deliberate symbolism, and effortlessly creates a continuity of sorts with the child's home environment. This includes floor sitting, a demonstration of poverty in unwashed walls and mats, a noise level and disorder that is more benign than a modernist disciplining, and the absence of any modern artifacts or images, including a common sense of the very impossibility of them. In the novel *mai*, we are told: "Dada's sitting room had been converted into a drawing room. His two portraits and sword in a leather sheath were still on the wall. But around them were English landscapes framed by babu in golden frames. Apart from this drawing room there were pictures on the wall only in the puja room. *We had not developed a tradition of hanging things on walls with a view to decorating the place*" (Shree, 2000: 99; italics mine).

This lack of wall decoration is a closely held ideal in the madrasa, and something that matches exactly the lifestyle of every student and teacher. I must not be misunderstood as suggesting that the madrasa is not a 'modern' institution. In all those aspects which characterize a modern education: age-graded classes, pre-determined and centralized curricula, examinations and report cards, the dominance of the textbook and the irrelevance of any particular teacher, it is as modern as all the other educational institutions which proclaim themselves as that. It is also modern in the historical Indian way that I have mentioned: in spite of a cultural-spatial continuity with the home, the madrasa school, like all schools, has a strained relationship with guardians. Each party in the relationship, the school and the home, blames the other for the inadequate socialization of the child into the self-conscious, confident, participating citizen of the modern nation state.

When we go beyond the symbolism of the spaces and look

at the curricula and rituals, all the schools, from madrasa to missionary present a set of patterns. The schooled child learns early on that the name of her country is *India*, that it is *her* country, and that it is *great*. The child sings songs to that effect, maybe marches or parades or salutes the flag, and takes part in celebrations on Independence Day and Republic Day. By class 3 or 4 the child studies about the thirty-two different states of the Indian Union, the boundaries, physical and economic features of each, and a bit about the costumes and festivals. Apart from the two all-India Boards that are popular in all towns, there are separate State Boards, which give more attention to the geography, and history of the particular state the school is located in.

My argument, however, is that the pedagogy of the nation is flawed. The efforts of all those who seek to create a new Indian citizen in the school have historically been lacking in technology, both its execution and its very conception. Their politics, for nationalists, have been correct - we must build up consciousness of and loyalty towards the nation - but their pedagogy has been insufficiently developed. Consequently, what has actually been offered to the child in the classroom has not been of the standard to produce the results desired. The schooled child in India has not developed that consciousness of the nation or that loyalty which was supposed to be produced. Although this argument could be demonstrated by citing the assessments of educators themselves, I have tried to look into explanations for it as well. Through ethnography in the classrooms, I show elsewhere that children in madrasa, private school, and public school, all experience a distancing from their history and geography that is a product of poor pedagogy. They cannot envision their country, cannot draw it, cannot refer visually to its physical features, cannot relate to its characteristics, or conceptualize any continuities - even ironic or subversive between it, the nation, and themselves in their experienced, lived-in lives. Through interaction and interviews with children I found that even those children who had mastered the arts of answering adult questions and spinning narratives within test procedures ("Name the five largest rivers of India"; "Where does it rain the most?") showed no ability to construct connections between their own biographies and their nation's history (Kumar, 2001).

If adults in India have not thought of pedagogic strategies that could successfully teach the nation, how do they react when they are directly *taught* the strategies, as in a workshop for the interactive teaching of nationalist history, such as I witnessed?

There were several reactions. The first was of agreement, a confirmation of knowing the same approaches, perhaps of having attended several workshops, of having taken training courses and got degrees. There was an agreement which seemed duplicitous, because there was a slight masking in the eyes and lack of volubility about precise work. There was an emphasized separation into 'theory' and 'practice' and the response was saying, "These are great ideas in theory *but not in practice*. I know the theory quite well, thank you, but I am clever enough to also know that practice is different." If we translate the discourse, it runs as follows. *Of course India is a nation, and a great, ancient one, the best in the world. But we Indians are too clever to fuss around with that. The real challenge to us lies in our confrontation with the problems of everyday life. Let us go on with resolving those. That is the practical. All this rhetoric about the nation etcetera is fine, but is only theory. (Not that the nation is theory, but to talk about it in this solemn, quasi-religious way will never happen with us, and is thus theoretical).*

Education Outside the School: Here Comes the Nation

When the child does not go to school, there are two main ways in which she encounters the nation, which are shared by school-going children too. One is in movies and the other is in songs, typically heard over radios, television, or over loudspeakers amplifying their sounds generously in streets and neighbourhoods. In contrast to the movies of the fifties and sixties, there are fewer nationalist plots today that elaborate on citizens' duties towards national development: the building of roads, dams, and high structures. That was perhaps not 'realism', but only idealism. Setting up a city in the jungle is a long-standing cultural ideal and was transferred from the king to the government. Movies of the nineties onwards are continuous with the older ones in their evocation of the natural beauties of the land: mountains, deserts, and waterfalls. But they are new in their direction of responsibility, not towards development, but to the protection of the virtue of the land and of its borders. Enemies are all those located outside: Pakistan which would seek to plant terrorism and attack the borders; and Europe, America, Australia and New Zealand, which are enticingly beautiful just like India but fail in virtue when compared with India. In movies such as *Pardes, Dilwale Dulhaniya Le Jayenge, Jeans, Ham Dil De Chuke Sanam, Pyar to Hona hi Tha, Kabhi Khushi Kabhi Gham,* India is not celebrated through its nature, but with a new confidence, through its value.

In this, the media, movies, television, and taped music, are all more advanced in nationalist rhetoric than school textbooks, which still continue to equate nation to nature. Textbook poetry asks: "What does 'nature' teach us?" and then answers with reference to the different parts of the national imaginary: "The Himalayas teach us to stand firm. The rivers teach us to nourish. The trees teach us to protect. The clouds teach us to share...."

Which is so much less resonating than the dialogue from one of the above films: "Are you a German?" "No." "Are you an American?" "No." "Are you a Russian?" "No." "Then?" "I am *an Indian*." The latter speaker has decided to fight for the girl's virtue, in contrast to the other nationalities evoked that presumably would have failed to. There are numerous variations of this theme.

The Neighbourhood: Garbage and Beauty

A trip through the *mohallas* or neighbourhoods of the city, along its main roads as well as its several lanes, inside the campuses and compounds of its main institutions and then a very close investigation of the official, commercial, residential, and recreational areas shows us that Banaras is a modern city with a particularly unified character. That is, not that it is not recognizably divided into inner and outer, upper and lower class, ethnic or caste, and educationally and occupationally separated areas. It is that the divisions, in so far as they exist, are secondary to other unifying characteristics, which, in the interest of brevity, are the following, illustrated by ethnographic glimpses.

When I watched fiom my balcony at seven a.m., I saw five or six women winding their way to and from between the houses of this colony. They were maids who worked in different houses, typically as cleaners, to do the sweeping, mopping, washing of dishes, and occasionally clothes. I focused on one woman who was just leaving a house. She must have been one of the 5.00 a.m. workers. She was carrying a packet of garbage in her hands. I knew exactly what would happen. She walked to the empty plot one house away in front of our house and aimed unenthusiastically towards the ground within. She missed, as do most women, and men. The garbage landed on the roadside, to join the growing piles on both sides, most of which had spilled out from its containers and blown and spread around. The twenty or thirty yards of the road at that part were a garbage dump itself.

Similarly, I watched and talked, with this woman and that, including our landlady, our neighbour, our maid, miscellaneous workers and loiterers, and confirmed that *all* of them took it as a law of nature, god, and society that garbage should be thrown wherever no one personally stood as occupant of the land. It is the first unifying fact about the neighbourhoods of the city. The difference that could arise, say out of regionalism, is slight, and subverted by other processes.

Vishvajit Chatterji, in class 9 of Anglo-Bengali Inter College, lives with his family in an ancient, crumbling house in a tiny *gali*. The house was in the corner, so they had some open space around their first floor rooms, but there were other equally tall or taller and older, houses a few yards away, so the room upstairs was dark and closed, in spite of its windows and balconies. More than half the room was taken up by a double bed, on which most of the social and educational activities were carried on. On the other side there was a fridge, a steel almirah and a dressing table. Visvajit and his brother Surajit cannot help noticing that their home looks different to their non-Bengali friends' homes, and similar to their Bengali friends'. Other points that might strike them as part of their Bengali identity are that their mother chops and cuts with a *boti*, not a knife. She cooks rice all the time, and fish regularly, as well as several other dishes in a Bengali way. They speak Bengali at home, and the children seem as fluent in Bengali as in Hindi. They celebrate Durga Puja in a big way. They go to Calcutta once a year and have relatives there. The mother's friends are almost all Bengali, although the father, being also a BJP member, moves across all sectarian lines. Where they are no different to all their neighbours is in their respect for the city's ethical codes about garbage, and its relationship to freedom. This is not articulated by them (as much of the above is not) and can only be observed. Father and son share the city's ethos, which in a nutshell, is the positive proposition: the more garbage, the more freedom, and *vice versa.*

Second, there is no discernible separation into public and private spheres of activities. That which may be considered a private activity at one place, for instance, bathing or massage, is explicitly a public one at another, and *vice versa*. That is to say, the code for public and private does not exist, and the city's ethos is unified in that it holds *one model* of the public and private: the lack of interest in a fixed division into public and private.

There is, therefore, no specialization of spaces by activities. Inside the house, cooking, eating, resting, and sleeping, gives place to working or meeting, without prejudice to any principle. Outside, every space from side of street to centre of street to park or empty public land may be used interchangeably for work, celebration, meeting, storage, and so on. There is perfect unity of social code and social practice.

Third, some arguably modern values, or more precisely, metropolitan modern values, are recognizably absent, including; the observance of rules of precedence in traffic, recognition of symbols for discipline, obedience to appointed law-makers, and a sense of that internalized control that would preclude, for instance, throwing garbage on any public space.

It has been my argument (Kumar, 2001a) that the nature of schooling, across the spectrum of all kinds of schools, is such that the discipline often boasted about or alluded to positively by schools and educators is not the discipline of Foucault's argument, that permeates the self and transcends conflict and consciousness, that empowers even as it controls, and makes the disciplined person a true subject of modernity. Rather the child in a provincial school learns only to be subjected to authority, and this authority is arbitrary both in composition and actual action, and arbitrary in its deployment of symbols. Thus, when free of scrutiny, the lessons of the school such as punctuality, straight lines, silence, cleanliness, or maintenance of codes of discipline, are totally unobserved by him. They are a kind of game that must be played only if those in authority are watching. This is not the phenomenon described by Paul Willis of working class kids reproducing their class through resisting a certain socialization (Willis, 1977). Rather, the children in a provincial city collaborate with the adult authorities to put on a show of learning discipline, and both collaborate in a kind of humanity that precludes learning the discipline, and in terms of the child's experience actually empowers him - within the provincial world.

But there is more to the teachings of the neighbourhood, and my fourth point is about the aesthetics of everyday life. A religious institution called the Gayatri Shaktipeeth had a grand celebration lasting for three days. It is one of hundreds of such sectarian public organizations, and they all have their periodic celebrations. The Shaktipeeth had dozens of loudspeakers spread over four neighbourhoods. They had two full days of chanting, lectures, yoga, meditation, *puja*, sermons, *prasad*, and singing. The theme was "Mahila jagriti" or the arousal of women. The speakers explained how important it was for women to rise and take up their full share of building up the society and the nation. The nation was taken to be homogeneous, with no difference of caste, class, or creed, as the cliché goes. The event was crowded. The arrangements were expensive and efficient. The sermons were articulate and powerful. The singing was tiresome.

In the field of a local school, a tent was erected where the public sat, women, or the sisters, on one side, and the men, or the brothers, on another. There were stalls of power-building *chyavanprash*, of incense and *yantras* and rings, of pictures of the founding teachers, and of books on the topics of the day. The stage on one side had six young women in saffron, playing, singing, and preaching. The audience was absorbed and often clapped or chanted alongside. At the edges children ran around playing their games.

What did these children learn from this event? Regardless of its specialized nature, it is one with large-scale feasts and rituals like weddings. Children could certainly catch odd things from the mikes on *mahakal*, on *shakti*, on the problems of contemporary society and their solutions. But what else? I would like to interpolate here an observation based on myself as an informant.

When I was small I often hung around till late at night at such events, sometimes falling asleep on the sofa or *dari*, running around with cousins if at home, with utter strangers if elsewhere. I always caught something of what was going on. I saw Indian dances first at such occasions, heard non-film Indian music first on such stages through such loudspeakers, and built up involuntarily my intellectual world with knowledge such as: what was so distinct about each of the Pandava brothers? Why did Mira hate her husband's family? How did Yashoda react to her son's naughtiness?

I learnt these things because I saw them repeatedly and not just once. But even more, I learnt how such events should 'feel', and the correct, normal ambience in them. If as an adult I came to arrange any large-scale ritual, it did not feel successful if it did not feel like that. Like what? With a tent (or marquee, as Salman Rushdie calls it) in the night, food served outdoors, people coming and going, amplified lights and sounds, and a sense of liminality. A oneness with nature, a continuity of space, an endlessness of time, the outdoors transformed into the indoors, but so much more exciting than the everyday indoors. These were things I learnt that were as concrete as the other.

The Home: Class

The sensual, symbolic, and metaphoric world of the provincial child learnt by her in the neighbourhood is shared across gender and class. By sensual I mean experiences such as those of the monsoons, of eating mangoes, of swimming in the river, and of playing in the streets. The symbolic is a continuation of the same, in that it constitutes and does not merely denote: the food, colour, sound, and liminality of festive occasions as necessary for both ritual correctness and the good life. By metaphoric, I mean the almost classical-by-now metaphors of the *ghazal* and the cinematic world, such as the cluster around love: stars, night, thirsty eyes, sleeplessness, tossing in bed, crying, calling out a name, responding. This world of the senses and metaphors is learnt also in the home, simply because the home is continuous with the neighbourhood.

The home of Gulistan and Gulfesa, ten and six was surrounded by a sea of water one day in August. The only reference I heard to the water was that the two girls would not be going for their tuition that day because of the water. Perhaps that was why they were happily playing cards instead. Gulistan and Gulfesa do not experience shortage of space or difficulty in any job in home because they have grown up in this home and they are totally used to it. This is how space and time are, as normal as sun is sun and water is water.

When I was small and visited my ancestral house in Kutchery Road, Lucknow, I never thought of its surroundings as a crowded street with open gutters on either side, and food shops which were made on wooden benches on top of these, the refuse from these shops going directly into the gutters underneath. There was a Thakur who brought around his cart every evening, ringing his bell as a special signal, and we would rush, if we could afford it, to slurp up his shaved ice and cream and syrup, never imagining that it was less than sanitary. The *only* consideration, ever, was, can I afford it? Only when I was grown up did I hear this beloved Kutchery Road described as 'dirty' and children being warned not to eat the '*nali ki chat*' (snacks of the gutters) or hang around by the roadsides. Moreover, all my cousins in this Kutchery Road house slept in smaller, darker bedrooms than did I in my cantonment bungalow, but I did not think their situation pitiable. Rather, it seemed wonderful and enviable. The eyes I have now to discern clean and dirty, spacious and crowded, well-lit and dingy, are not the eyes that children ever have, no matter who they are.

When I was small, the rainy season was a time of magic. Clouded skies, breezes, drizzles and sprinkles, a kind of magic quality in the air. There was a chance of 'rainy days' and school being unexpectedly closed. It was possible to sit in the verandah and work all afternoon, without being bothered by heat or cold. A child, after all, does not have to be conscious of adult monsoon problems like mildew and mould. I remember many play kitchens, with little clay or metal cooking utensils, leaf and clay foods, picnics, real and make-believe. I remember swinging on the *charpoy* swing, or being safe under a roof, soft greens outside.

Maybe Gulistan and Gulfesa are enjoying their childhood monsoons equally. The delight on their faces certainly seems to bear witness to it. This, despite the fact that their house is broken down, very cramped, surrounded by filthy water, and they are themselves resource-less, poorly fed and nightmarishly educated.

Of course class is important. The toilets of poor people are a problem - for the outsider, but may be for them too. Some are totally outside the home, in the worse cases being the side of the nearby pond. Little children in poor localities are encouraged to defecate on the wayside, and almost all children to urinate anywhere. When slightly better, homes have toilets in the courtyards, entrance ways, and staircases. They are dark, with non-flushable commodes in the ground, of the squatting 'Indian style', with stored water to be dispensed with a mug. The child learns to manoeuvre among all this matter-of-factly, and never actively suffers.

Bathing places are not specialized either. The tap or well is not in a closed space. Men and women bathe routinely in courtyards, with some items of clothing on. In other cases, there is no private well or tap and buckets are brought in from outside. There is a preferred place for keeping them, but obviously a bucket could be kept anywhere. Thus a child could one day be bathed on the roof and another day in the corner of a room. The Ganga is a vast reservoir for one's private use, including brushing of teeth, washing of clothes, and soaping and scrubbing oneself clean.

The night spaces in the cramped rooms get messy and are slowly transformed. A thorough *jharu* (sweeping) is done on all the floors, by the mother or an elder sister, never by the father. Sheets and blankets are folded up and stacked away. Mattresses and *chatais* are rolled up and put away. Some beds are stood up; others are left flat for daytime rest. Where there are permanently placed beds they serve as sofas, couches, diwans, and are just like the floor in bed-less places. The floor everywhere is made clean enough to sit and eat at. Nowhere do people wear shoes into the house or living areas.

But class socialization is much weaker than gender socialization. Unlike with gender where both the parents are simultaneously present for comparison, there is no other model before the child except that of the home she is born into. Those whom she visits are other members of the family or close friends, and are typically of the same class. Or she assimilates, as in my childhood visits to the ancestral house, two different lifestyles as equally 'natural'. Only the adult suffers pragmatically from shortcomings, and only the outsider interprets spaces, processes etc. to be limited or cramped. For the child they are normalcy itself and she finds them not remarkable in the least.

The House of Bachhe Lal, a Painter, and the Process of Gendering

I will suggest that all children do *not* share the same experience of space, nor express themselves similarly with regard to it, but that the line of divisions are not those that are the strong, evident ones of education and class. Neither the difference between the schooled and unschooled child, or the child in elite public school *versus* the child in the madrasa; nor the difference between the elite and lower class child, if both products of the provincial city, is the crucial axis of division. Children are, on the national plane, divided according to whether they belong to metropolitan or provincial cities, towns, and villages, because the nature of neighbourhood and community life and teaching differs in each case. Within the provincial city, the experience of space and the expression of space vary according to the sex of the child.

In the house of Bachhe Lal, a potter, there live his mother and father, and his five brothers, of whom two are married and have seven children among them. The house is an open space with three separate rooms or huts, all *kachcha*, of mud. The open space is like the water in a map of the earth with some land interspersed in between. On one side is usually spread a curtain on which Bachche Lal paints typical studio scene of buildings and roads with artificial flowers sprouting in the foreground. It is very iconical of the much discussed modernist tropes of the road leading to the new India (Srivastava 1998). On the other side lie some masks drying, in symmetrical rows of deer, bears, monkeys, lions, and so on.

The treatment of children in this family was a puzzle. Again and again I was assured that nothing of the family craft was taught to them until they were much older, and that even then they just picked it up from watching. The children seemed to have little to do, with no toys (Bachche Lal chose to interpret my question about toys as the masks and dolls made in the family and shrugged about how satiated the children were with those) and nothing child-size in the house that I could see. They were not allowed to touch any of the materials. There were of course no separate spaces for children at all. It seemed self-defeating not to provide them with anything of their own and then scold them incessantly for trespassing into the adults' worlds. They would stand around, feel the pressure of bottled-up energy and surge forward to touch something or act upon something, and be shooed or struck out of the invaded space again. Adults were forever creating a centre from which children were barred and of which they were supposed to stay in the margins.

Bachche Lal had actually made a children's landscape that was ordered, purchased, and now hung up in the neighbourhood school. He took me to see it. *It was exactly like the landscapes he made for his photo studio clients*. There were mountains in the background, a stream in front with boats on it, a bridge on one side, trees and a settlement with huts on the other. There were bright colours, a sunset, flowers, shadows. There were no people. It was abstract and distance and frozen in time.

On Diwali day, it felt that the activities going on were no different to other days. The house was not cleaned up or decorated. The family behaved like the producers of the objects being celebrated—images, clay lamps and toys—with attention to others' consumption in the market rather than on their own consumption. In the middle of my mid-morning visit, they packed up two crates with these things, ready to go. One was picked up by the youngest brother, a boy of about eight, and one by the father. They headed for Khojwa market, where two other sons had set up Diwali stalls. I followed. The little boy walked fastest, and led the procession. Khojwa, even after hundreds of visits, can remain a mystery with its lanes and byways, all suffused with garbage. The last time I had to find a place, I was also led by a boy of maybe ten, who led me all along back lanes in the interest of saving time and showed me another world than I had know existed, this one with even more garbage.

Then I remembered other incidents from my fieldwork. I remembered all the scores of times I had been sent to wherever I needed to go under the guidance of little boys. Or, when I had asked directions at a shop or on the road and a little boy had stepped forward with alacrity to guide me, and adults had taken it as a given that he should. Then other consistent patterns: little boys of even four or five sent out for *pan* and tea to the corner shops. Women's voices echoing throughout the day inside homes, calling out to sons, nephews and little brothers to fetch them this or that from the market, from outside.

The Outside

Before beginning an analysis of the inside-outside division one must repeat that in India there is no exclusive space for children, of whatever age. When they are small they are not taken anywhere specially by their mothers/parents, as, for instance, to a park but to places where everyone goes, such as markets, or others' homes, or the ghats. When they are independent, from just three or four onwards, they wander into each other's homes, congregating wherever convenient, say on rooftops, or in the lanes themselves. When slightly older, they seem to take over vacant lots, neighbourhood squares, and burial grounds, but only temporarily, because they can be and are chased out regularly from these places. The same spaces are used for work, such as drying yarn, selling or demonstrating wares, and for adult recreation, ranging from the most informal to the most formal. Then, when they are adolescents, the children have absolutely nowhere to go. Embarrassed to play in open places, they congregate in tea or pan shops, on ghats and streets.

At the early age of six or seven, the category of 'children' gets divided into 'boys' and 'girls'. The Hindi word for children, *bachhe*, tends to be used thereafter only for boys, and thus have I used 'children' above. The separation is on many counts: clothes, speech, body language, activities, and even food. But its most dramatic dimension is in its location of children: in the 'inside' for the girl and the 'outside' for the boy.

At the times of the day when school begins and ends, both girls and boys will be visible on the streets, coming from or going to school. Girls walk openly, or cycle, or ride rickshaws, scooters, or buses, apparently in equal numbers to boys. Sometimes they walk in groups and seem to take over parts of the street like boys do. The difference lies in where this takes them, if I might be forgiven this pun of mobility. Apart from travelling to and from school, mobility is not a structural characteristic for girls. They do not go out for pleasure, for hanging out, or to pass the time. There would be nowhere to go. A street side, *pan* shop, tea shop, or any open urban space is not the provenance of the female of any age, unless they are professionals or otherwise have work. From early morning to late night, females are visible on streets as sweepers, garbage collectors, vegetable sellers, and hawkers, and of course, shoppers in transit. But they do not stand around without work, as men and boys do.

The little boy in Banaras gets the training to be a flaneur early on in his life. The city is his oyster, starting from his neighbourhood. One could write extensively on the subject of men's activities in the city (Kumar 1988); what would remain to be added is that the socialization into this occurs early on for children. There is scanty literature on this, even as for girls there is the proposition that in colonial times they came to belong to the inside, to stand for the home, purity, and the non-colonizable nation. At this point I can only give tiny ethnographic insights regarding the gendered ways of the working of space.

All children, for example, wake up to the sound of their parents. Gender socialization begins at that point, because the mother is typically filling water, doing the dishes, or sweeping the floor, and soon lighting the fire for the first meal. The father is either still asleep, or also busy. In some cases his work entails going before dawn to buy the raw materials for his business, such as *pan*, vegetables, or milk. In some cases, he has to start setting up his shop by carrying things out. In other cases, he might spend the first few hours of the day to grooming himself and preparing for the day, whether with a newspaper and cup of tea, or a leisurely toilet, or preparing what he needs for the day at home. *In no case* is the father the person who is responsible for the morning meal and the children getting fed and sent off with their tiffin boxes.

The only exceptions are when the mother is ill, dead, or absent because of an emergency somewhere. Then the father does take over, ready to pass on the job to the oldest female. An occasional informant, such as Lajjavati, might narrate how her father brought her up completely after her mother's death and how he remains the best cook she ever knew. All the women of *Assi*, my research neighbourhood, get a welcome break from cooking when they are ill or having their periods. As their daughters grow up, however, such as Lajjavati's Chunni, which means when she is as little as nine or ten, the daughter takes over the substitution work from the father.

The mechanisms of all this are clear enough to the child, and the gender socialization is stern and unalterable. The gender socialization, as with class, seems natural and is not remarked upon. Almost all children take for granted that the mother should provide all the services at home, as well as are around the whole day like an anchor for the various vessels to come back to and tie themselves up temporarily. They do not consciously think that the woman's situation could be compared to the man's or any constructive ideas generated from the comparison.

But there may be more commonalities than we imagine between the sexes too. I will end with one woman's and one man's song. The first is a *vidai geet* or the lament of parting from her family by the bride as she leaves for her new home. The second is the man lost in the world, ready to wander. Uprooted as they both are, could it be that the pain is so great only because they would both *like* to be fixed in a place?

The woman sings:

> *.... sakhi sakhiyan bachpan ka ye angna guriya jhule koyi bhi to hoga sabnam*
> *chhupaungi aansu kaise bhigege kangna*
> *sakhi sakhiyan re....*
>
> *...my friends, my girlfriends, the courtyard of my childhood, the dolls, the swings....*
> *how will I hide my tears [as I depart]?—my wedding bracelets will get soaked...*
> *my friends, my girlfriends....*

The man sings:

> *ai dil mujhe aisi jagah le chal jahan koyi na ho*
> *apna paraya meherban ya meherban koyi na ho...*
> *duniya mujhe dhunde magar mera nishan koyi na ho*
>
> *ah, my heart, take me to a place where there is no one*
> *there is no one of mine, no one other, no friend or supporter at all...*

where the world would keep searching for me but there is no sign of me...

REFERENCES

Abu-Lughod, L. 1993. *Writing Women's Worlds: Bedouin Stories*. Berkeley: University of California Press.

Bourdieu, P. 1977. *Outline of a Theory of Practice*. Cambridge: Cambridge University Press.

____. 1990 [1970] "The Kabyle house or the world reversed", in *The Logic of Practice*. Oxford: Polity.

De Certeau, M. 1984. *The practice of everyday life*. Berkeley: University of California Press.

Donald, J. 1996. "The Citizen and the Man About Town", (pp. 170-190) in S. Hall and P. du Gay, (eds.), *Questions of Cultural Identity*. London: Sage.

Fabian, S.M. 1992. *Space-Time of the Bororo of Brazil*. Gainesville: University of Florida Press.

Foucault, M. 1975. *Discipline and punish: the birth of the prison*. New York: Vintage Books.

Fuchs, D. 1995. "Training and Continued Education—Children's Culture", *European Education Summer*, 26(2): 5-17.

Giroux, H.A. 1983. "Theories of reproduction and resistance in the new sociology of education: critical analysis", *Harvard Educational Review*, 33: 257-93.

Holston, J. 1989. *The modernist city: an anthropological critique of Brasilia*. Chicago: University of Chicago Press.

Kumar, N. 2000. *Lessons from Schools: The History of Education in Banaras*. Delhi and Thousand Oaks, California: Sage.

____. 2001. "Children and the partition: History for citizenship", (pp. 269-301) in Suvir Kaul (ed.), *The Partitions of Memory: The Afterlife of the Division of India*. Delhi: Permanent Black.

____. 2001a. "Learning modernity? The Technologies of Education in India", *Les jeunes: hantise de l'espace public dans les societies du Sud? Autrepart,* 18, pp. 85-100.

____. 2002. "Why do Hindus and Muslims Fight? Children and History in India", in D.P. Mines and S. Lamb (eds.), *Everyday Life in South Asia*. Bloomington, Indianapolis: Indiana University Press.

Littlejohn, J. 1967 [1960]. "The Temne House", (pp.) in J. Middleton (ed.), *Myth and Cosmos: Readings in Mythology and Symbolism*. New York: American Museum of Natural History.

Low, S.M. 1999. "Spatializing culture: the social production and social construction of public space in Costa Rica", (pp. 111-137) in S.M. Low (ed.), *Theorizing the city: the new urban anthropology reader*. New Brunswick, New Jersey: Rutgers University Press.

Moore, H.L. 1986. *Space, text and gender: an anthropological study of the Marakwet of Kenya*. Cambridge: Cambridge University Press.

Rabinow, P. 1982. "Ordinance, discipline, regulation: some reflections on urbanism", *Humanities in Society*, 5: 267-78.

Rabinow, P. (ed.). 1984. *The Foucault Reader*. New York: Pantheon Books.

Redfield, R. and Singer, M. 1954. "The cultural role of cities", *Economic Development and Cultural Change*, 3: 53-73.

Shree, G. 2000. *Mai: a novel*. Translated from Hindi by Nita Kumar with an afterword. Delhi: Kali for Women.

Srivastava, S. 1998. *Constructing post-colonial India: National character and the Doon School*. New York: Routledge.

Willis, P. 1977. *Learning to Labour: How Working Class Kids get Working Class Jobs*. Aldershot, U.K.: Grover.

Chapter 7

IN BETWEEN TWO CULTURES

The Process of Educational Institutionalization of Tribal Children in Orissa, India

Deepak Kumar Behera

The educational institutionalization of childhood refers to a process by which organized arrangements, chiefly the school system, influence children's lives and organize their days. Schools are specialized institutional settings dedicated to instructional purposes in which adult experts concern them exclusively with the education of the child, and communicate with them verbally in a room set aside for that purpose. When children in a community have previously had no schools to begin to attend, their lives become different from those who never went to school. They become subject to an authority outside the family and are increasingly connected with a world with different values and ideologies outside their community. *Such children are placed in between two cultures.*

When one examines the social context of learning in communities without schools, many features become clear. In such communities, children learn to master adaptive tasks, but there are no special institutions or settings dedicated to instruction. They learn through participation in economic production and other activities organized by adults, beginning as peripheral participants who perform simple and repetitive tasks while having observational access to the mature practice of their elders. It may be that no one pays attention except when they do something wrong; they receive guidance only as a corrective feedback. Gradually they progress to more advanced tasks they have seen others perform. In many tribal societies where economic production is at home, the process can begin in the third or fourth year of life and continue up to adolescence. One would see such children doing exceedingly well outside the framework of formal schooling. Primitive education was a process by which continuity was maintained between parents and children, even if the actual teacher was not a parent but a - maternal uncle or shaman. Modern education includes a heavy emphasis upon the function of education to create discontinuities to turn the child of a peasant into clerk, of the farmer into lawyer, of the illiterate into literate (Mead, 1942: 633-39).

Against this backdrop, an attempt has been made in this paper to examine the process of educational institutionalization of the tribal children of Ashram School in the State of Orissa. The study analyses the impact of residential schooling system on the life of the tribal children. It reports the daily life of the tribal children in the school and the hostel settings.

The Concept of the Ashram School

The concept of *'ashram'* is based upon the Hindu philosophy. Traditionally the *'ashram'* was a centre of learning where the teacher (*guru*) and the students (*shisyas*) were living together as part of the institutional complex. One major objective of this system was to provide the learners with a congenial environment to improve their skills and craftsmanship under the able guidance and supervision of the *guru*.

Following the recommendation of Dhebar Commission, a number of Ashram Schools[1] were opened up in India during the fifth five-year plan (1970-75). Initially such residential schools were introduced in various Tribal Sub-Plan (TPS) areas for the educational advancement of tribal children. An Ashram School admits students from Standard I to VII and provides them with residential facilities. The school also admits some day-scholar students living within a radius of 3 km. from the school. Such schools are being managed by the Welfare Department of the State Government. Sometimes an Ashram School may be upgraded to a high school with teaching facility from Standard I to X[2]. Even then it continues to function under the same Welfare Department of the State Government.

Tribal Literacy Scenario in Orissa

There are as many as 62 tribal communities in Orissa. The percentage of tribal population in the State was 22.21 per cent as against the India's percentage of 8.08 per cent[3] (1991 Census). The medium of instruction for school children remains the same through out the State; the tribal children are not exception to this. According to 1991 Census, the literacy rate among the Scheduled Tribes in India was 29.60 per cent (40.65% for male and 18.19% for female). The literacy scenario was still worse in tribal Orissa; it was 22.31 per cent for the entire Scheduled Tribe population of the State (34.44% for male and 10.21% for female)[4] as per the same Census. The formal education/literacy programme has not attracted the tribals of Orissa to a great extent because they do not find it life-oriented.

The Sample Schools

All the three Ashram schools selected for the study were located in the Tribal Sub-Plan areas of Orissa. They are, namely, Gonasika High School, Balipeta High School and Singsari High School. Gonasika High School is located in Bansapal Block of Keonjhar district. The school has teaching facility from Class VI to X. The tribal students of the school are from Juang, Gond, Saora, Santhal and Bhuiyan backgrounds. There are 13 teachers in the school. Baliapet High School is located in Narayanpatna Block of Koraput district. The school admits students from Standard IV to X. It provides residential facilities to 200 boarders. There are 13 teachers (out of the total sanctioned post of 22) in the school. Some of the hostel rooms have been converted into staff quarters due to shortage of space. Majority of the tribal students are from Kondh and Paroja communities. Singsari High School is located in Umerkote block of Nabrangpur district. The school has classes from VI to X with boarders' strength of 120. A total number of nine teachers (out of the total sanctioned post of 15) serve in this school. Majority of the tribal students are from Kondh, Gond, Bhatra and Poraja backgrounds.

Tribal Children in the Sample Schools

Typically, children's life in institutions are analysed from the point of view of the latter's socialising or educative capacities or functions as caretakers: do they actually meet the requirements for the schooling of children (Qvortrup, 1994: 8). Kovarik (1994: 103) uses the concepts of "stage" and "script" as spatial and temporal characteristic of the life world of children: the stage refers to the designing and physical structuring of children's environment, while the script is the agenda for how children are supposed to spend their time within a given stage. The time used by tribal children in Ashram school is largely determined both by a stage and a script - limited by economic factors and directed by adults. Ashram schools have institutionalized the separation of the tribal students from their society, confirming upon them a separate identity: their proper place is the school and the hostel. The process of educational institutionalization constructs tribal children out of their community, mutes their voices and denies their personhood.

For the tribal students of the Ashram school, the dominating feature of childhood is that of powerlessness and lack of control over what happens to them. The activities of these children at school are bureaucratized by the adults (the teachers). Children are the objects of the school enterprise, persons to whom actions are done. As the traditional tribal socialization is non-authoritarian, the tribal children dislike the threatening environment in the school. A tribal child would not submit to any physical beating, shouting or insult. Those students who somehow rather manage to continue for sometime fail to secure the expected grade and eventually feel humiliated. Institutionally peripheralised childhood restricts tribal children's activities to "child-specific places". The activities of the students of such schools are strictly scheduled in time. Further, each kind of activity has its own adult leadership.

The students find it monotonous to sit continuously in a similar posture without talking to their classmate for a long six hours. Sitting in an over-congested classroom with only little a break-time makes them feel completely exhausted and tired. They are exposed to too-extensive textbooks, excessively complex subject-matter, unintelligible formulations, a host of scientific terms, abstract concepts, encyclopaedic facts in textbooks etc. (Kovarik, 1994: 113). The mass of teaching material is disproportionate to the age of tribal students and the degree of difficulty of textbooks rises irregularly and disproportionately. Some of the most criticized features of the schools are: regimentation; a rigid system of seating, grouping, grading, and marking; and the authoritarian role of the teacher (Toffler, 1987). The urban middle class oriented education is superimposed on the tribals both in terms of its structure and content (Sharma, 1976).

The textbooks used in the sample Ashram schools remain the same as in plain areas. These books are printed in Oriya language. The medium of instruction in the school is also Oriya. Majorities of these tribal communities have their own dialects, which substantially differ from Oriya (the officially recognized language of the State). Though it has been accepted in principle that primary education should be imparted in mother tongue, this remains to be translated in realities. Therefore, it becomes difficult for the tribal children to comprehend the subject teaching along with language learning (Barla, 1965). The phenotypes of a particular language can hardly be correctly translated into a different language. Lack of comprehension in case of children leads to dropout, stagnation and truancy, as they do not find the urban middle-class oriented education interesting.

The curriculum does not reflect tribal values and ideologies. It does not take into consideration the socio-cultural environment of the area and the people. The tribal students hardly find anything interesting and life-oriented in the curriculum (Chottopadhyay, 1953). They read about people and things that are unknown to them and their culture. The tribal inmates of the sample Ashram schools are made to learn and understand the alien realities, which are not at all of concern to them, e.g. the picture of a scientist in a sophisticated laboratory or a mother wearing jeans. They are taught physics, chemistry, geometry and geography instead of hunting, fishing, ethnomedicine and agriculture.

In the absence of sufficient teaching aids in these schools the teachers fail to communicate to the students effectively. Therefore, the tribal children take education as a burden. They find the course structure too top heavy for comprehension. Some of the students stated that they find it difficult to understand the ornamental language used by the teacher. A few others are not satisfied with the explanation given by the teacher on a particular topic. The students find it again difficult to understand things, which are very abstract by nature.

Many of the senior students stated that they believed the existing course curriculum was not suitable for them and it needed revision in the light of the tribal culture. The students also expressed their dissatisfaction about the disbursement of scholarship, supply of text books and reading materials, availability of teaching aids, supply of sporting materials, etc. For these reasons they say they try to escape from it at anytime and by any means. This attitude leads to dropout and stagnation. This also affects their general proficiency and achievement in the examinations. The special coaching that is given to the students preparing for standard X examination does not yield the expected result. Coaching is given to these students in English, Mathematics, Science and Sanskrit. Each school receives only a small amount of funds, from Rs. 400 to Rs. 600 per annum for this purpose. Due to the lack of funding the teachers who are engaged in this special coaching don't take the task seriously.

Sending a grown-up child to school for a tribal family is a matter of economics and it creates dislocation in the traditional pattern of division of labour. Many parents just cannot afford to send their children to school. As agricultural activities of the tribal are family-oriented, assistance of children is very much required, especially during certain periods of the year. The parents, therefore, withdraw their children and engage them either in farm or wage earning activities. Again the traditional tribal festivals draw away many students from the Ashram schools during certain period of year. The students are given holidays to observe Ganesh Puja and Saraswati Puja (both are caste Hindu festivals) and *not* tribal festivals like Karma, Kandul, Chait, etc. The common educational holidays calendar of the Government of Orissa followed in all schools, including the Ashram school, does not take into account festivals and agricultural activities of the tribal communities. Many tribal students are first generation learners and therefore, their parents take little interest in the activities of the school. There is a weak student-teacher-parent continuum in the school.

Tribal parents view the timely marriage of their daughters as an important aspect of their tradition. As a result, female students of Kanyashrams (not included in our sample) drop out from the schools as soon as they reach puberty. Many of them also drop out in order to perform household chores and take care of their younger siblings. The parents believe that "too much education will spoil their daughters". Eventually it would be difficult for them to find out suitable grooms for their daughters. If the expected economic returns are not forthcoming the parental support, while initially quite high, drops precipitously after a five- to eight-year period. The parents expect that the student, who completes his schooling, would eventually land a white-collar job, and would remit money regularly. These expectations are increasingly unfulfilled, as the white-collar workforce has become saturated in the State of Orissa. This has now created in the minds of many tribal parents a negative attitude to formal education.

The vast majority of teachers of our sample Ashram schools are from non-tribal background. They can neither speak nor understand the local tribal dialect (Bardhan, 1973; Kar, 1962). Some of them are sent to these Ashram schools on punishment transfer. They do not have even the working knowledge on tribal values and ideologies. Many of them do not stay with their families in the school because of the lack of infrastructure facilities and the general backwardness of the region. Therefore, they remain homebound and are not generally found at the place of their duty. One would find them busy in pulling strings to get them transferred to urban places. Consequently, there is little communication between teachers and students. When a non-tribal teacher finds it difficult to understand the tribal dialect, how can it be expected that a tribal student would acquire expertise in a language other than his mother tongue? Being out of tune with tribal culture and language, the teachers possess a negative attitude towards the tribal students. They perceive the tribal students as 'not-yet human beings', 'not-yet responsible' and 'not-yet knowing'. The act of judging tribal children 'incompetent' by the teachers severely restricts their autonomy to consent to or refuse matters affecting their lives. The rigour of a strict disciplinary regimen, long arduous hours and a work pattern dictated by the school system leave no space for the tribal child to participate in decision-making processes. The children never reach maturity or independence at school. The independence that teachers say they aim for in children turns out to be conformity with school norms, both academic and social. Teachers impart teaching and training to the children within the framework of their understanding of the 'good child'.

The inmates have to make *shrmadhan* (in the language of the teachers) during their leisure. The teachers consistently engage the inmates in their personal work, viz. cooking, sweeping, cleaning utensils, fetching water and firewood, gardening, marketing, taking care of the children, etc. Further, students are asked to assist the cook in preparing the food. The students are also engaged in maintaining the school garden where vegetables are normally grown. It is a pity to note that the school teachers take away a major part of the vegetables that come out of the school garden. Afraid of being punished, the children perform all the above-mentioned activities without any protest, thinking those to be well-accepted items under the concept of *shrmadhan*. Even a small mistake/protest by the student is dealt with by severe punishment. The inmates discuss among themselves matters such as the control exercised by adults over their free time; how their daily lives are organized and controlled by teachers; and their opportunities for activities and social groupings outside the immediate control of their teachers.

Another major problem in the sample schools is the lack of strong interpersonal relationships between students and teachers. The students do not feel free to discuss their problems, viz. classroom teaching, course content and examinations, hostel problems, etc. with the teachers. The teachers also usually do not encourage the students to do so. It is significant to mention that all the teachers of the sample schools follow a non-participatory method of teaching. A 'monologue classroom situation' is a common sight. As a social group, the students see themselves as actors whose position required them to negotiate with and occasionally fight against the teachers. Most of the teachers are involved in local politics and engaged in petty business. For many teaching is a secondary occupation.

Tribal Children in the Hostel

Residential facilities for the inmates of the Ashram schools are extremely poor. The classrooms get converted into sleeping rooms during night hours. The inmates lead a life under sub-human condition and remain half-starved in the hostel. They are provided with two main meals in a day. The menu remains the same throughout the year excepting on some special days when the school organizes students' feast. Students are normally served with rice and *dalma* (when vegetables are cooked together with gram, it is called *dalma*). The majority of the inmates suffer from nutritional deficiencies. One would hardly find any medical facility in these schools. Only in extreme cases are the inmates taken to hospital. Having suffered from chronic diseases for prolonged periods, many inmates leave the school permanently. The school does not conduct periodic health check-ups for the inmates. The toilet and bathing facilities in the school are far from satisfactory.

A fairly large number of inmates (sometimes even 25 to 30) are given accommodation in a single room (30ft. × 20ft.). The rooms are ill maintained and therefore, one finds them in really bad shape. There are no proper lighting or ventilation facilities. The students sleep on the floor. The rooms get flooded with water during the rainy seasons because of the constant leakage from the roof, door and window. The inmates find it difficult during that time even to locate a place for sleeping/resting. The majority of the inmates expressed their dissatisfaction about accommodation, quality of food, bathing facility, toilet facility, lighting facility, medical facility, recreational facility and supply of essential articles.

The inmates get up early in the morning. A bell is rung to awake them in time. All of them report for a group prayer at 5 p.m. An attendance call is taken and the students who are found absent during the prayer are given severe punishment. After the prayer, they get a break time of half-an-hour to attend to nature's call. One would find the school latrines and bathrooms in really bad shape; thus the majority of the students prefer open air for this purpose. On their return the students are engaged in activities, viz. cleaning the school premises, gardening, cleaning the utensils used for cooking in the previous night, assisting the cook in preparing the lunch etc. Some of them go to their teachers' house to perform their household chores. The students are asked to read and write for an hour or so. The students are then made free to take a bath and get ready for the school. The food (rice and *dalma*) is served to the students around 10 a.m. The students then attend classes between 10.30 a.m. to 4.00 p.m. with two breaks (one for 10 minutes after 2nd period and the other for 30 minutes after 4th period). On their return from the school they are given one-hour time to play. Then they almost repeat the same work as done by them in the morning session (cleaning the school premises, gardening, reporting in their teachers' house for household work, assisting the cook in preparing the dinner, etc.). Again there will be an evening prayer and all the inmates attend the same. After the prayer, the students and teachers sit for some time and the duty chart for the students (prepared by the teachers) will be read out for the information of the students. The students are supposed to report their problems to the teachers in that meeting. However, out of fear they do not bring their problems to the notice of the teachers. The teachers also do not encourage the students to make such complaints. This is followed by a study hour for about 2 hours. The superintendent and some of the concerned teachers are expected to visit the hostel during this time and help the students in their studies. However, the teachers rarely visit the hostel during that period, and the inmates don't get enough opportunities to clarify their doubts. Then a bell rings indicating that it is now time for dinner. The students are again served with rice and *dalma.* After that the students relax for some time and then retire to bed. Some of the responsible and senior students are given some portfolios by the teachers.

In each of our sample Ashram schools there are a number of student ministers, viz. *Mukhya mantri* (the chief minister), *Swasthya mantri* (the health minister), *Shrunkhala mantri* (the discipline minister), *Sanskrutika mantri* (the cultural minister), *Bagicha mantri* (the garden minister), *Kastha mantri* (the fuel minister), *Saphaee mantri* (the cleaning minister) etc. In one Ashram school, I had the privilege of interacting with two such ministers, viz. the health minister and the fuel minister. The health minister (a student of class X) had with him an old wooden box containing some bottles filled in with liquid and some tablets. The minister had an instruction manual with him indicating the use of the medicines contained in the box. The superintendent had instructed him to tackle all health related problems during odd hours without disturbing him. So whenever someone has an ailment, he reports the matter to the health minister and the latter gives him the medicine as per the instruction manual. Each medicine in the box had a number and the instruction manual states the use of a medicine bearing a particular number. The health minister only remembers the number of the medicine without knowing their names. He gives medicine to an ailing student seeing the number of the medicine. Most of the tablets in the box had passed the expiry date. The health minister reported to me that when he finds it extremely difficult to handle a case, he reports the matter to the superintendent. The experience of the fuel minister is equally revealing. Every alternate date he, along with his friends, goes to the nearby forest for collecting firewood. They come back to the school carrying a lot of firewood. Although such collection is illegal, the teachers turn a blind eye to this. Half of the collection goes to the hostel mess. The remaining half is supplied to the teachers, by the fuel minister, for their personal use. The teachers misappropriate the money that the school receives for the fuel consumption of the students.

Educational Incentive

The educational incentives meant for the tribal children do not reach them in time. There is considerable delay in the sanction and disbursement of scholarship. The quantum of scholarship (Rs. 300 per month, i.e. a little less than US $6) provided to each inmate is found to be inadequate due to the rapid escalation of prices in the recent years. The small amount received by the school for each student covers their food expenses, daily needs, washing allowance, medical expenses, etc. The following table gives a break up of the sanctioned amount of Rs. 300.

Table 8.1: Showing the item-wise break-up of the monthly scholarship received by students

Item	*Quantity per meal*	*No. of meals*	*Total quantity*	*Price per kg.*	*Amount per day*	*Monthly expenditure*
Rice	250 grms	2	500 grms	Rs.8.00	Rs.4.00	Rs.120.00
Dal (Gram)	25 grms	2	50 grms	Rs.30.00	Rs.1.50	Rs.45.00
Vegetable	100 grms	2	200 grms	Rs.5.00	Rs.1.00	Rs.30.00
Oil	7.5 grms	2	15 grms	Rs.50	Rs.0.75	Rs.22.50
Fuel and salt	–	–	–	–	–	Rs.20.50
Fish/Meat/Egg		4 times in a month	–	F-Rs.50M-Rs.100E-Rs.1.50 each	Rs.5.00 each time	Rs.20.00
Total						**Rs.258.00**
Kerosine						
Electricity					Rs.0.30	Rs.9.00
Soap					Rs.0.30	Rs.9.00
Hair/body oil					Rs.0.15	Rs.4.50
Medicine						Rs.2.50
Dress						Rs.17.00
Total						**Rs.42.00**
Grand Total						**Rs.300.00**

This amount is sanctioned to the inmates for ten months (and not for 12 months) in a year. Further, the amount is not released every month; rather it is sanctioned in three to four instalments in a year. Frequently the government sends a circular to all the Ashram schools/Kanyashrams stating that the next instalment cannot be released for at least the next four to five months due to financial constraint. The headmaster is supposed to open an account in the name of each inmate and deposit the amount in his/her account as soon as he receives the instalment. However, this is not followed in any of the Ashram schools. The headmaster finds it extremely difficult to manage the situation because of the non-timely receipt of the instalment. He purchases the majority of the items on credit basis from the local shops. He buys fewer quantities of items with the plea of the non-receipt of money in time. When the headmaster receives the instalment, he repays back the credit and he often misappropriates the remaining amount along with some school teachers. The headmaster tackles the situation by providing chicken and liquor to the inspecting staff during their visit to the school.

Each inmate of the Ashram school is entitled to receive free of cost a blanket, bed sheet, mosquito net, utensil, cut, bucket, mug, etc. from the school. However, it is a pity to note that most of the inmates do not receive these items due to their short supply by the government. Therefore, the teachers of the school take away the limited supplied items (meant for the inmates) to their houses for their own use.

The following statements reflect the attitude and perception of the tribal students of the sample Ashram schools towards their school and hostel environments:

> "Our teachers occasionally visit our hostel and share with us our problems beyond school hours". *Student of standard IX, Balipeta High School*
>
> "I find it difficult to concentrate in my studies, as I feel hungry during major part of the day". *Student of standard IX, Balipeta High School*
>
> "I do not like the way we are put to task by our teachers. Frequently the teachers engage us in their personal work". *Student of standard IX, Singsari High School*

> "I am really fed up with the stereotype food provided to us during lunch and dinner. Those really do not taste good".
> *Student of standard X, Gonasika High School*

Concluding Remarks

If modern educational theory is stressing the development of independent and creative young minds - and if traditional tribal education implies the same development - why should tribal students of today be exposed to an educational system that was devised 100 years ago in Victorian colonial times and is thoroughly authoritarian?

Many partially educated tribal children who have dropped out from the school, fold, or completed a certain level of education, are unwilling to work in the fields and unable to find jobs, eager to wear 'shirt/pant'. One will find them dancing to the tune of a silly Hindi film song and not according to their beautiful melodious tribal folk song. Such cases - widely spread - demand explanations.

The concept of 'child' is culture-bound. As the tribal child moves from its family and community to the Ashram school, she or he slowly gets uprooted from the cultural moorings of his community. The new and alien school environment filled in with *civitas* values and ideologies sets new standards of behaviour and norms for the tribal children. Placed in such a peripheralised environment the tribal students develop a sense of alienation. Thus, the process of educational institutionalization, by keeping the tribal children away from their family and community environments, has further marginalized childhood. Schooling for many tribal children acts neither as a channel for upward mobility nor as an instrument of personal development. They reside in a cognitively limited environment. The price they pay is what they do not become, what they cannot enjoy, what they fail to comprehend. Having completed the process of educational institutionalization when the tribal child permanently returns to its family and community, s/he finds himself caught *in between two cultures*, i.e. the *civitas* and the tribal. The school-returned child could be compared with *a bird with broken wings*. The childhood in tribal society has been invaded by rules and classification schemes based on alien values and cultural standards. To sum up, the process of educational institutionalization through the Ashram school has created a kind of colonization of the life of the tribal children of Orissa.

NOTES

1. The parallel residential schools meant for tribal girls are known as "Kanyashrams". Such schools have not been covered in our sample.
2. Sometimes such schools may have teaching facility from Standard IV to X or Standard VI to X or Standard VIII to X.
3. Registrar General of India, Decennial Census, 1991, Series 1, Papers 1 & 2, Final population totals.
4. Registrar General of India, Decennial Census, 1991, Series 1, Paper 1 of 1993, Primary Census Abstract for SC and ST.

REFERENCES

Bardhan, A.B. 1973. *The Unsolved Tribal Problem*. New Delhi: Communist Party Publication.

Barla, A. 1965. "Medium of Instruction for Tribal Students," *Indian Education*, 4: 11-15.

Chattopadhya, K.P. 1953. "Tribal Education," *Man in India*, 33(1): 27-31.

Kar, B.C. 1962. "The Role of Teacher in Tribal Community," *Journal of the University of Gawhati*, 13a Arts: 15-19.

Kovarik, J. 1994. "The Space and Time of Children at the Interface of Psychology and Sociology," (pp.101-122) in J. Qvortrup, M. Bardy, G. Sgritta, H. Wintersberger (eds.), *Childhood Matters: Social Theory, Practice and Politics*. England: Avebury.

Mead, Margaret. 1942. "Our Educational Emphases in Primitive Perspectives," *American Journal of Sociology*, 48: 633-39.

Qvortrup, J. 1994. "Childhood Matters: An Introduction," (pp.1-23) in J. Qvortrup, M. Bardy, G. Sgritta, H. Wintersberger (eds.), *Childhood Matters: Social Theory, Practice and Politics*. England: Avebury.

Sharma, B.D. 1976. "Planning for Educational Development in Tribal Areas," Occasional Paper on Tribal Development-II, New Delhi: Ministry of Home Affairs.

Toffler, A. 1987. *Future Shock*. New York: Random House.

Chapter 8

CONTENTMENT AND COMPETENCE

Rajasthani Children Talk about Work, Play and School

Ann Grodzins Gold and *Bhoju Ram Gujar*

The late Sharon Stephens suggests in the introduction to her groundbreaking volume, *Children and the Politics of Culture*: "children creatively live from the inside complex mixtures of languages and social domains that are external structures for many adults" (1995: 24). We take this to mean that children through their experiences know something, or many things, that adults do not know or fail to grasp - because adults' experiences are compartmentalized while children's are not just hybrid, but fluid. More recently, Lawrence Hirschfeld in an article provocatively titled, "Why Don't Anthropologists Like Children?" makes a similar if more radical claim in somewhat different terms: "Children not only live in the cultural spheres of the adults with whom they share a life space ... but they create and maintain cultural environments of their own" (2002: 615). Where Stephens sees "complex mixtures" inaccessible to adults, Hirschfeld suggests we may locate unique cultures created by children.

In his call for "mainstreaming" the anthropology of children and childhood, Hirschfeld, ironically, is interested in those cultural environments "created and maintained" by children largely because they may yield "significant insights about the nature of adult cultural experience" (2002: 611). However valid that quest for revelation may be, our wish both in our research and in composing this essay is not to examine Rajasthani children in order to learn more about either the adult world or the human mind. Rather, we hope modestly to unfold a few interrelated aspects of children's perceptions and appraisals of their experiences. Here we find ourselves in accord with Scheper-Hughes who states strongly, "As long as children are valued in research (as in life) only for what they are ultimately bound to become (i.e., non-children), childhood is diminished..." (1998:14).

The findings presented here emerge from our collaborative attempts to learn about childhood in rural North India through methods that are loosely ethnographic. At the time of our main research endeavour in 1997, Gold was a grey-haired American anthropologist in her early fifties; Gujar was a Rajasthani educator in his early forties. The two of us have collaborated since 1980 on many projects, but our research on children and the transmission of knowledge was the first of our collaborations to emerge directly from Gujar's expertise and interests.

Gujar uses his caste name as a surname; he is a member of the Gujar community, a group traditionally associated with herding and dairying although just as likely today to practice agriculture. He is the most educated person in the local Gujar community, and was at the time of our fieldwork a respected middle-school teacher, addressed as "Mar-sa" (a Rajasthani adaptation of the British "master").[1] We began talking to children experimentally in 1993, gathering their environmental perceptions in whatever haphazard manner we could (Gold and Gujar, 1994). Fascinated by what we had learned, we conceived a research proposal to study the transmission of environmental knowledge both informally among herding children, and in government schools where environment had recently been added as a subject to the required science curriculum. Thus our central concerns were with processes of learning in two broadly contrasting settings. These were homes, fields and grazing lands where transmission of knowledge was oral, contextualized, and practical; and primary and middle school classrooms where teachers strive to follow a set curriculum and most teaching is text-based, generalized, and abstract.[2]

Who are the children to whom we listened and from whom we learned? All of the herding children came either from middle-ranking communities whose primary occupations are farming and herding, especially Gujar, Mali and Mina; or from the more disadvantaged Regar and Chamar, formerly leather working castes whose primary livelihoods today are also agricultural. All these herding children were from Ghatiyali, the large village in Ajmer district that is Gujar's birthplace and home, and that has been Gold's primary research site since 1979.[3] Although our central focus in this article is on herders, we also draw on our work with school children during the same research period. Many of these children were from in and around Palasiya and Gandher - two villages in which Gujar has taught, both located in Bhilwara District where he has been posted since the nineteen-eighties. The majority of the school children with whom we spoke were also from non-elite groups, including Gujar, Dhakar and Bhat. These children were, for their families, pioneers in the uncharted world of literacy. About twenty per cent of the school interviewees were from Brahmin, landlord or merchant communities - representing a higher social class, and often having literate parents. Children from these communities may help with livestock and do other agricultural chores on weekends and holidays, but none were represented among full-time herding children.

As our project developed in the field, our specific focus on the transmission of environmental knowledge diffused. We attempted to grasp more holistically the radically different but overlapping and intermeshed worlds of school children and herding children. Gujar himself had worked at herding in his childhood and youth. Thus, although he is at present an adult authority firmly grounded in the world of literacy and degrees, his personal experiences bridged the worlds of school and herding in ways that made him wonderfully able to talk to children in both locations. A foreigner, Gold spoke Hindi with an accent; although she dressed in Indian clothes, her uncovered head made her appear very different from all Rajasthani women of her age. Children generally found her more amusing and baffling than threatening or authoritative.

Our field methods throughout the research period were eclectic. Adults cannot exactly operate as participant observers among children. Failing that, we did attempt to be attentive, sympathetic, non-intrusive companions and listeners. In schools, Gold often took a passive observer's role. She sat long hours in classrooms, schoolyards and headmasters' offices - just listening, watching, and writing notes. She recorded performances at numerous school assemblies. Sometimes, with their teachers' cooperation, she was able to take a small group of children to walk outside the school to identify local plants, making lists of their names and uses. As a teacher, Gujar had often to shift between a research scholar's role and his daily duties as an active agent in the situations being researched.

Fieldwork with herding children included following them and their animals out into the grazing lands surrounding the village to see how they passed the long days (see Fig. 8.1). As one girl put it, the animals "graze on their own," affording children much opportunity for play. The animals are often quite far away from their herders; it strikes us that none of our photographs show children and livestock in the same space. Nonetheless, even young herding children seem able, casually but responsibly, to keep just enough of their attention focused on their charges while enthusiastically engaged in other activities.

We watched herders in a wide variety of play activities from sedentary to highly active. These included, to give just a few examples, intense concentration on "board games" scratched onto rocks with pebbles for markers; the imaginative construction of mud "houses" (Fig. 8.1); athletic exertions in the game of *kalam dali* or "stick-branch" which unites tree-climbing with a kind of tag. While the play house was definitely a girls' project, none of the other games we saw were gender segregated. It was charming to watch herd girls deftly fold and tuck their long skirts to form a neat and serviceable pair of "half-pants" before launching themselves into the trees (see Fig. 8.2).

Gujar, easily able to draw on his own childhood experiences, pleased these children, put them at their ease (and enjoyed himself) by joining in their games or helping them stack their firewood bundles.

Each time Gold stepped into a schoolyard she recoiled inwardly from the palpable tension that pervaded the atmosphere - at moments instilled further by a military veneer, as at the regimented

Fig. 8.1: Herd girls pose beside their mud play house

Fig. 8.2 : Herd girl up a tree, in a lively tag game called "stick-branch"

daily morning assemblies (see Fig. 8.3). It seemed to Gold that Gujar himself, while playing his "master" role, sometimes

Fig. 8.3 : School children stand at "attention" during daily morning assembly

manifested a different personality as he participated in the school's relentless commitment to altering persons.[4] He sometimes spoke sharply with children when they were unresponsive to our questions. But even when he was gentle and coaxing - as was normally the case – children often remained disturbingly tense. Although pressed to sit down during an interview, many children stubbornly remained respectfully standing (see Fig. 8.4).

Gold found herself contriving to avoid interviewing school children, and her field notes express some of this strain:

> *Bhoju agrees that the mode of speaking with, interacting with, children [both his own and other adults'] is very different in the schools; that only those boys and girls who are very strongly motivated (he says) or brave (I say) can keep going. The shy and hesitant will shrivel and wilt. I would have become a goatherd myself if I were a girl in Ghatiyali. But, how to summarize the differences and the*

Fig. 8.4 : Bhoju Ram Gujar interviews anxious-to-please school boy.

> *gulfs? At school if a child makes a mistake he or she is chastised, ridiculed; to speak one must stand up and risk all (yet I admit a few eager hands do wave). At home, most of the time, children are cooed at;... The old folks who herd with the children barely "instruct" them;... they learn organically, they copy, they see and do, perhaps like lambs learning to graze... At school although teachers enforce discipline, they themselves are undisciplined; they wander, they are not punctual,... There is a general sense [in their attitudes toward the children] of "these peasants, what's the use!"*

We were perpetually struck with the contrast between the way herding children and school children acted in our presence. Children out of school, and especially out of their houses and away from senior kinspersons, were articulate, forward, carefree, and unashamed - boys and girls alike. After a chance encounter in mid-January with some girls gathering firewood, Gold's diary noted:

> *They have no fear of us, they talk about their work, they say they learn at home, from the people of their house; they say they don't go to school because they are afraid. They laugh happily and pay us very little mind and go about their work . . . I ask myself why these girls are so forthright, so un-self-conscious; their ages are probably between ten and fourteen. When we try to talk with girls of the same age in the setting of the school they are nearly paralyzed; they whisper, unable to tell us even the simplest things that they surely know, standing stiffly even when singing.*

Eventually Gold came to suspect her own revulsion from the school atmosphere as an overreaction based on miscomprehension. No enduring damage was done; the tension sustained at school dissolved easily just a short distance from the building where children's high spirits readily returned and even their body language changed. In the evenings school girls and herd girls played together happily in the village neighbourhoods (see Fig. 8.5).

Fig. 8.5 : Street game at night outside Bhoju Ram Gujar's house

Contentment as Competence

In our interviews we often asked children simply framed questions using a language of work and play - concepts with which adults readily compartmentalize various aspects of life both in English and Hindi. But as we reviewed the ways children would describe themselves and their occupations in the two contexts of school and herding, we saw that this language, although freely used by children in their answers to our probings, tended to be introduced by the interviewers. In the children's own responses another more important concept emerged. Herding children in particular evaluated their lives and activities in terms of whether or not they felt content or amused and capable. All of these experiences are wrapped up in a single intransitive Rajasthani verb: *asangno* (sometimes pronounced *asyangno*). This was the standard by which these children judged their own well-being.

Gujar's quick gloss on *asangno* was Hindi *man lagna* - an idiom whose literal sense is for the mind to be attached. Its semantics include to feel mentally comfortable or to feel at home. The Rajasthani-Hindi dictionary (Lalas, 1962-1978) also gives *man lagna* to translate *asangno* along with other phrases that encompass a semantics of enjoyment and competence. It is telling that essential ingredients of enjoyment and contentment are familiarity and empowerment (for the Sanskrit root from which *asangno* derives is a term for power).[5] Thus, we stress it is not just pleasure, perhaps not even primarily pleasure, but autonomy and above all competence that children gauge when they talk about situations that do, and do not, yield a positive experience of contentment. In transcribed interview texts, we translate the verb *asangno* variously, as "to feel content" or "to enjoy oneself" (see Fig. 8.6).

It was mostly unschooled herders who used this word, and they used it frequently, to describe what they liked about herding; and in the negative, what they did not like about housework (this from girls) and knew they would not like about school (from both boys and girls). School children in the upper grades, who have adjusted to school routines, may describe education favourably, in

Fig. 8.6 : Three herding children at their ease

contrast to herding and agriculture, as an existence characterized by rest and comfort (*aram*). The herders' life is physically strenuous (or vital and exuberant in the case of these generally healthy children). It has many opportunities for varieties of enjoyment and diversion, but is surely never considered a life of rest and comfort. Just as surely, the school-goer's life is not one that fosters a sense of competence and relaxed contentment (*asang*); we see this in the looks on their faces (see Fig. 8.7).

Fig. 8.7 : School children seated

Conversations with Herding Children

All of the interviews with herders cited here were done in late January, early February 1997 over a ten-day period when the power was out in the village due to a transformer breakdown. No computer work could be done at home, so to take whole days in the fields with the children was particularly attractive to us. The weather was lovely; crops were growing in irrigated fields, but non-irrigated agricultural land was open to grazing, making herders' lives easier.

One day we encountered a group of young *Malis* (a community whose traditional work is vegetable gardening). An older woman named Mangi Bai who was available to help them when they needed help, or advise them when they encountered any problems, accompanied these boys and girls. She jokingly called herself their "Mar-sa" or teacher. We had a long conversation with one smart and talkative boy among this group named Ramesh. He said he had no idea of his age, but we place him in the ten to twelve-year-old group by guesswork.

Bhoju : *Did you ever go to school or have you always grazed cows and buffaloes?*

Ramesh : *I never went to school. In school one isn't content (skul men asyanga kauna).*

Bhoju : *In what ways are you content with cows and buffaloes?*

Ramesh : *We play and we run and we don't have to sit in one place.*

Bhoju : *What do you have to do in school?*

Ramesh : *If we go to school we have to read/study and the masters beat us; the masters beat us and the other boys also beat us.*[6]

Bhoju : *What other work do they have you do at school?*

Ramesh : *The masters demand water, and they make you clean up.*

Bhoju : *And?*

Ramesh : *And they make you sweep.*

Bhoju : *Did you ever go to school?*

Ramesh : *Never at all.*

Bhoju : *How do you know then?*

Ramesh : *My brother goes to school - my younger brother, he goes.*

Bhoju : *So what kinds of things do you do, for the whole day long?*

Ramesh : *Nothing we graze the cows and sit around.*

Bhoju : *What else do you do?*

Ramesh : *We play with marbles, and sit.*

Bhoju : *How else do you amuse yourselves while herding cows and buffaloes?*

Ramesh : *We run here and there and play, and if roj [nilgay, "blue bull"; a kind of antelope] come then we chase them away, and we play with marbles, and if we see titar [a kind of small bird, snared and cooked by herding children] then we run to kill them. In the hot season they can't fly fast and after chasing them a while we can catch them. And now [in this season] we get honey from the beehives. We are looking for them now in the kair bushes.*

Having found such a satisfyingly talkative young man, we continued to ply him with questions. He told us about songs and stories that the herding children share, and gave many examples. He talked about what to do when your charges ran away. He produced a very long list of trees and shrubs and further awed us with his knowledge of herbal remedies for infections and rashes. Bhoju asked him if he would go to school, should his folks decide to send him; he clicked his mouth in the vehement Rajasthani negative and pronounced: "I wouldn't go!"

Ramesh's interview paints a pretty complete picture of the good life as conceived by a young herder. Work and play are united seamlessly in this portrait of well-being. Ramesh feels himself not subject to any other wills except those of the animals over which he exercises fair control. Mangi Bai, the older woman who joked with us about her "teacher" role, is evidently a resource in time of need, but does not represent any kind of superior or disciplinary authority. Ramesh's contentment includes not only many pleasant diversions, but a strong sense of his own command of knowledge and skills, of competence in daily life. All of this contrasts strongly with the alien space of school where children are told to forget all they ever knew from their home lives and challenged to acquire an entirely different set of practices (Gold, 2002).[7]

Gulabi, a small Mali girl, is in charge of one buffalo, one female buffalo calf and one female goat. Although most unschooled girls and boys will refuse, like Ramesh, even to guess their own ages, Gulabi tells us confidently that she is about ten years old. She is one of four sisters and she also has two younger brothers. One brother goes to school in the first grade, and the other is still too young. Gulabi rejects the notion of school and praises the gender-free herder's life in contrast to the female realm of housework as well.

Bhoju : *Did you ever go to study?*
Gulabi : *No.*
Bhoju : *What do you do at home?*
Gulabi : *I dump the cowdung, I get the water; I bring drinking water from the Sweet Well. [There is no public water tap in the Mali's neighbourhood.]*
Bhoju : *How many times a day do you go to get water?*
Gulabi : *Twice a day, and I do the sweeping.*
Bhoju : *What other work do you do in the day?*
Gulabi : *I graze the goats and bring home firewood.*
Bhoju : *Do your folks tell you to do it, or do you do it on your own?*
Gulabi : *I do it on my own.*
Bhoju : *If you didn't bring wood would your mother scold you?*
Gulabi : *No...*
Bhoju : *Do you want to go to school?*
Gulabi : *No.*
Bhoju : *What work do you like?*

Gulabi : *I like grazing livestock.*
Bhoju : *Why not housework?*
Gulabi : *At the house I am not content (asyangta nahin) and the flies bite.*
Bhoju : *But herding, isn't it a lot of trouble? The sun beats down, you have no water, and all that?*
Gulabi : *Even so I am content.*
Bhoju : *Who taught you how to do this work?*
Gulabi : *Nobody taught me, they graze on their own.*

Gulabi's vision of contentment stresses the well-being she feels outdoors, away from the congested village where bodies, foods and refuse attract flies, especially in this winter season. She also prefers grazing, it seems, as a space with considerably less supervision from adults. Another herd girl, whose name did not get recorded, also said she had never attended school. She was equally ready to dismiss any thought of going there. We guessed her age also to be about ten years. When Bhoju asked her about her work, she countered with a narrative of play.

Bhoju : *What work do you do at home?*
Herd girl : *Nothing. I play with marbles and then I eat and go out to graze.*
Bhoju : *With whom do you go grazing?*
Herd girl : *With all of them, these ones [she gestures at her companions].*
Bhoju : *And if you are hungry?*
Herd girl : *Then I bring bread from there to here.*
Bhoju : *And if you are thirsty?*
Herd girl : *Then we go one by one, or two by two, to the well. In the morning we give them [the livestock] water and in the evening on the way home. [The children can then drink when their charges drink.]*
Bhoju : *What do you play?*
Herd girl : *Marbles, and if there is water then we bathe, and if there is mud and water we play kyara-dhora ["plant*

beds-water channels" - a gâme where children use mud to model garden beds and irrigate them, as their elders do in this winter growing season, by building small barricades which channel water flow].

Bhoju : *If they sent you to school would you go?*

Herd girl [emphatically] : No!

Bhoju quizzed this child on the names of trees and birds, and she knew a great many; she also knew about the plants that the cows must not eat, and said she was always careful that her playing did not interfere with her care for the animals entrusted to her. Herding may be fun, but it is not a game. All the herding children are well aware of the value of their work in terms of domestic economy, although not all stress it in interviews.[8]

One day we met up with Bhoju's own nieces: Kamalesh, his sister's daughter and Kali, his mother's brother's daughter. They are young teens sparkling with mischief. When we came upon them they had just successfully raided a beehive and their hands were covered with honey and beeswax that they were futilely attempting to wipe off in the dirt, as there was no water source at hand. They looked messy and a little sheepish, but were clearly happy. They spoke of their herding lives as spaces where it is always possible to be content (*asyang*); if one diversion isn't available, they readily find another.

The girls told us with great glee of a dangerous form of amusement practiced by *badmash* (naughty) girls - possibly themselves. Sometimes these naughty girls take wet sticks and throw them up on the electrical wire, and it makes sparks, and they have fun. Once the electricity man came on his bicycle and asked who was doing this, and they said "not us!" Nonetheless, he took all their wood away from them. It seemed that their play had caused problems with the line running into the village, blowing several fuses. But the girls, unrepentant, emphasized that there was great bliss (*anand* - a more powerful word than contentment) to be had from watching the sparks. Sometimes, they confided, when the green wood landed so that it crossed two wires, enough sparks fell to earth to allow them to ignite a small fire, its warmth welcome in this winter season when the mornings and late afternoons could be a bit chilly.

A Mina girl, with the unusual Sanskritic name of Paremeshvari, contrasted the pleasures of herding to whatever she imagined she would encounter at school. She is proud of her housework and her herding work, expressing her sense of accomplishment in both areas.

Bhoju : *Did you ever go to school?*
Paremeshvari : *[a negative click]*
Bhoju : *Did your family not send you or didn't you go?*
Paremeshvari : *I myself didn't want to go!*
Bhoju : *Is it good to go to school or stay with the animals?*
Paremeshvari : *With the animals is great contentment (accho asang hai) - because we play "stick-branch" and gather firewood, and keep on playing, so we enjoy ourselves!*
Bhoju : *What do you do at home?*
Paremeshvari : *I take care of my little brother, and sometimes I make food. I wash the dishes, make tea, and collect cowdung - that's the kind of work I do!*
Bhoju : *What work did your mother teach you?*
Paremeshvari : *How to roll bread.*
Bhoju : *Did your father teach you any work?*
Paremeshvari : *He taught me to take care of sheep and goats (ber bakri ka kam); how to give milk to the lambs; and how to tie up some of them. If the animals come hungry from the jungle then I have to give them some green fodder, and I do that work too, getting fodder (see Fig. 8.8).*

It seems to us that for Paremeshvari as well as other herding children, life yields satisfaction, or contentment, because it is full of things they know how to do well. Familiar herding is often contrasted with the unknown but fearfully imagined school. For

Fig. 8.8 : Herd girl with small sickle used to cut leaves for fodder.

reasons of their own, it seems that school children will impart to herding children, even their own siblings, all the most negative rumours about school experiences. These may be exaggerations, but they also reflect genuine aspects of a deeply rooted reality. The contrast that emerges between grazing land and schoolrooms is not simply a contrast between a place of games or fun and a place of difficult study; nor is it merely a contrast between the acquisition of practical working skills *versus* the acquisition of abstracted knowledge. The fear of school that herding children express is a pointed fear of incompetence - or rather of the humiliation and punishment that incompetence would inevitably bring in the school's perceived alien realm.

We are aware, of course, that children's views are partial. Our article does not address, for example, larger economic structures that might well motivate adults to reinforce working children's fears of school, when their labours are needed for grazing animals. Our attempt here, to take seriously herding children's own expressions of why school is not an attractive place for them, is thus necessarily limited in scope. We are nonetheless convinced that Rajasthani children's voices and viewpoints deserve consideration by educators who want to improve the situation, as much as by anthropologists, who may only want to understand it. To return to Stephen's argument with which we opened, we have attended to the "complex mixtures of languages and social domains" those children live and articulate and thus learned about some crucial factors perpetuating gulfs between grazing and education.

Of course, some children without literacy in their family backgrounds do succeed in school, against the odds, and thereby achieve new realms of competence, new sources of satisfaction. Their numbers increase every year. Gujar himself was one such child. In his dedicated work as a teacher and headmaster, he tries to temper the discipline schools require with gentler encouragement and humour - especially for children from the farming and herding communities he knows so well. The herding children themselves made us understand clearly that only such tempering would allow more of the boys and girls from non-literate families to stop imagining school as a place where they will be made to feel perpetually incapable, and hence perpetually discontent.

NOTES

* The authors gratefully acknowledge the support of a major grant from the Spencer Foundation for the research upon which this paper is based. Oral versions were presented in November 2002 at the Annual meeting of the American Anthropological Association and in January 2003 at the National Institute of Advanced Studies, Bangalore, and we are thankful for many comments received that contributed to our thoughts on childhood. We are particularly indebted to Nita Kumar and A.R. Vasavi. Above all we thank the children from whom we learned so much.

1. Since his promotion in 1998, Bhoju Ram Gujar is a headmaster in Bhilwara District.
2. See Gold 2002a; Gold 2002b; Gujar 1997 for more on this project.
3. We refer readers who desire more ethnographic, historical and environmental context for the children's views presented here to Gold 1988 and Gold and Gujar 2002.
4. See Gold 2002b, Sarangapani 2003, Vasavi 2001 for much more on the walls between school culture and local culture in Rajasthan and elsewhere in India. For the most perceptive treatment I know of educational history in India see

N. Kumar 2000; N. Kumar 2002 offers reflections and insights with broad implications on the ways school children do and do not internalize historical knowledge of India's partition. See also K. Kumar 1991 on colonial construction of education and its negative legacies.

5. Lalas' Rajasthani-Hindi dictionary defines the intransitive verb *asangno* as *man lagna*; *dil bahalna* (to be amused or diverted); *sahas karna*; (to act bravely); *adhikar men karna* (to take something into one's authority) (1962-78). Gujar elaborates the semantics from an adult viewpoint as implying a situation that "causes you to feel happy" like having a TV or a son. These prime signs of success in the eyes of the world suggest that the term is as much about relations with others as experiences within the self.
6. See Sarangapani 2003: 148-150 for a description of threats and punishments in a rural North Indian school; the behaviours observed there are very similar to those we have seen in Rajasthan.
7. See Gold 2002b; see also Sarangapani who writes of a school in a large village just north of India's capital, Delhi, "The strong framing of the school curriculum is predicated on the belief that the present reality outside the school is not worthy of engagement" (2003: 261).
8. See, however, a conversation with Kali Gujar at about nine years of age, recorded by Gold in 1993. Kali, a herd girl, speaks of the value of herding in contrast to the uselessness of school. Asked by Gold, "Is grazing good or is studying good?" Kali answers definitively, "Grazing sheep is very good." Asked to elaborate why, Kali replied in no uncertain terms: "Sheep give us income, what does reading give?... They give lambs, so we sell them; they give dung, and we sell it" (Gold 2002b: 91).

REFERENCES

Gold, A.G. 1988. *Fruitful Journeys: The Ways of Rajasthani Pilgrims.* Berkeley: University of California Press.

Gold, A.G. 2002a. "Children and Trees in North India," *Worldviews: Environment, Culture, Religion,* 6(3): 276-299.

Gold, A.G. 2002b. "New Light in the House: Schooling Girls in Rural North India," (pp. 86-99) in D. Mines and S. Lamb (ed.), *Everyday Life in South Asia.* Bloomington: Indiana University Press.

Gold, A.G. and Gujar, B.R. 1994. "Drawing Pictures in the Dust: Rajasthani Children's Landscapes," *Childhood* 2: 73-91.

Gold, A.G. and Gujar, B.R. 2002. *In the Time of Trees and Sorrows: Nature, Power and Memory in Rajasthan.* Durham: Duke University Press.

Gujar, B.R. 1997. *'Without Trees and Plants We Will All Die!': Learning Nature in Rajasthani Schools and Fields,* Unpublished paper, Syracuse University South Asia Center Seminar Series.

Hirschfeld, L.A. 2002. "Why Don't Anthropologists Like Children?", *American Anthropologist,* 104: 611-627.

Kumar, K. 1991. *Political Agenda of Education: A study of Colonialist and Nationalist Ideas.* New Delhi: Sage Publications.

Kumar, N. 2000. *Lessons from Schools: The History of Education in Banaras.* New Delhi: Sage Publications.

Kumar, N. 2002. "Children and the Partition," (pp. 269-301) in S. Kaul (ed.), *The Partitions of Memory: The Afterlife of the Division of India.* Bloomington: Indiana University Press.

Lalas, S. 1962-1978. Rajasthani Sabad Kos, 9 vols. Jodhpur: Rajasthani Shodh Sansthan.

Sarangapani, P. M. 2003. *Constructing School Knowledge: An Ethnography of Learning in an Indian Village.* New Delhi: Sage Publications.

Scheper-Hughes, N. and Sargent, C. 1998. "Introduction," (pp. 1-33) in N. Scheper-Hughes and C. Sargent (eds.), *Small Wars: The Cultural Politics of Childhood.* Berkeley: University of California Press.

Stephens, S. 1995. "Introduction: Children and the Politics of Culture in "Late Capitalism" (pp. 3-48) in S. Stephens (ed.), *Children and the Politics of Culture.* Princeton: Princeton University Press.

Vasavi, A.R. 2001. *Discrimination and Deprivation: Basic Education Among the Poor.* Bangalore: National Institute of Advanced Studies.

Chapter 9

CONSUMING SPACES

Growing up on an Island in the Pacific Ocean

Karen Malone

The global community's image of Pacific Islands is often one of paradise - azure water, palm trees, and other tourist brochure clichés. In reality, Pacific Small Island Developing States (SIDS) face serious and unique challenges that render them vulnerable and fragile, comparable often to the world's poorest nations. Like many postcolonial peoples, the Pacific islanders find themselves both exploited and marginalised by a global system of mass consumption. Mass tourism, global warming, overfishing, resource degradation, pollution and loss of culture are only some of the long-term impacts of globalisation on these fragile islands. These are islands where governments need a "no room for error" management strategy and yet most have a very short history of 'westernised' self-governance and limited internal political capacity to compete in a global economy. Life presents significant and unique challenges particularly for many children living on fragile and isolated Pacific Islands. There is a tension between their seemingly idyllic lifestyle on the one hand and the hungry tourist grazier, globalisation and a lack of good governance on the other. Informed by the work of social theorists who are attempting to understand the modern identity and its relation to space and time, this paper attempts to move beyond more traditional anthropological views of culture and space as secular and fixed, to a new perspective. A position where "the modern self is inextricably tied to fluidity or movement across space and time" (Rapport and Dawson, 1998 : 4).

The children's stories in this paper, presented as a number of short vignettes with images, have been collected as part of the UNESCO project Coastal and Small Islands (CSI). The CSI initiative seeks to enhance actions for sustainable living on small island nations and to nurture opportunities for collegial capacity building across regions and across oceans. Within CSI is the community-based project Small Island Voice (SIV). SIV seeks to bring to the table islanders' voices of how they are negotiating changes in their local environments to the wider global community. Working with the CSI-SIV team the author's role was to support children from the island nations to participate in a shared community voice. Participatory research methods based on the UNESCO *Growing Up In Cities* (GUIC) model were utilised in this process (Chawla, 2002; Driskell, 2002). The data collection methods included semi-structured interviews with children and other community members, surveys, child-led guided environmental tours, children's drawings and daily time schedules. Presented in this paper is research data collected with the primary school age children (5-12 years old) of two Pacific Island nations: Cook Islands and Papua New Guinea. Both rural (outer island) and city/urban locations have been presented. It will conclude with a discussion on environmental quality from the children's perspectives utilising the qualitative indicators of quality of life developed by the *Growing Up In Cities* project team (Chawla, 2002). I will begin the paper with a short review of relevant literature around identity and space, culture and globalisation, environmental sustainability and children's rights.

Reconceptualising Identity and Space

Hebdige (1990, cited in Urry, 1995) argues that a growing scepticism concerning older explanatory models of identity have led to a renewed interest in the relatively neglected, 'undertheorised' social dimensions of space. It is becoming less common in social and cultural theory for space to be represented as neutral or fixed. Instead space itself is being refigured as, "inhabited and heterogeneous, as a moving cluster of points of intersection for manifold axes of power which cannot be reduced to a unified plane or organised into a single narrative" (Hebdige, 1990 cited in Urry, 1995: 167). In response to this critique there is a need to appreciate the complexity of the modern experience and to examine the traditional anthropological interest of identifying and localising groups and societies and the tendency to represent cultural separateness rather than cultural and societal interconnectedness (Christensen *et al.*, 2000). Rapport and Dawson (1998) for example, argue against static conceptualisations of identity formation in relation to space and emphasise the 'compressing' of cultures and 'hybridisation' of identities accelerated by globalisation and synchronicity. Hybridity theorists such as Anzaldua (1999a), Bhabha (1994), Hall (1996), and Luke and Luke (1999) argue for a view of identity as hybrid, as constructed from multiple experiences and relationships that are enacted within particular spaces, places and time. This reconceptualisation of identity in relation to space suggests the modern self is inextricably tied to fluidity or movement across time and space. This view recognises that identities are neither fixed nor given; they are not ascribed or belonging to a particular culture or living in a particular space but the negotiation of multiple spaces and multiple cultures (Christensen *et al.*, 2000). A multiple view of identity and space informs the theoretical work underpinning this study.

Culture and Globalisation

The island is a powerful metaphor in anthropology and many other disciplines (such as biology). Isolation, uniqueness, boundedness of systems - the physical, social and cultural 'island' is frequently portrayed as the archetypical metaphor of a closed, self-sustaining social and spatial system. In so far as the local community described was compelled to have relations with the outside world, these relations would be depicted as extrasystemic links, as not really forming part of the relevant social unit. These views of societies, or cultures, as being closed social and spatial systems have come under scrutiny in post-modern literature where it has been stressed that no society is entirely isolate, that cultural and physical boundaries are never absolute and that the webs of communication, influences (whether direct or indirect) of a global economic and political markets have seeped into even the most isolated spaces. That is, cultures and societies are tied together in increasingly complex ways. A plethora of academic works exploring the relationship of globalisation and culture intersects these systems across the disciplines of technology, economics, sociology, cultural studies and anthropology. The boundaries become blurred. "Globalisation lies at the heart of modern culture, cultural practices lie at the heart of globalisation" (Tomlinson, 2001: 1).

Environmental Sustainability and Children's Rights

The Convention on the Rights of the Child (CRC) identifies the importance of providing safe and supportive environments for children. It outlines through a variety of Articles (2, 15, 17, 29, 30, 31) the importance of the child's right to play and access to recreation facilities, a capacity to participate in the cultural life of their community, their right to associate freely with others, access to information and the opportunity to participate in decisions that will impact on their well being. To support these rights member governments must guarantee protection from discrimination and violence. Additional to these rights the Preamble to Habitat 2 Agenda recognises the special needs of children in regard to the importance of providing opportunities for them to participate in processes that ensure sustainable human settlements:

> The needs of children and youth, particularly with regards to their living environment, have to be taken fully into account. Special attention needs to be paid to the participatory processes dealing with the shaping of cities, towns and neighbourhoods; this is in order to secure the living conditions of children and of youth and to make use of their insight, creativity and thoughts on the environment" (UNICEF 1997, Paragraph 13).

The principles of sustainable development clearly demand that the simultaneous achievement of environmental, social and economic goals meets the needs of the present generation without compromising those of future generations. Children have a special interest in these goals - they are the future contributors, decision-makers and citizens of the world. The goals of sustainability insist nation governments maintain the integrity of the social, economic and environment fabric of their global and local environment through processes which are participatory and equitable. The principles of the Convention on the Rights of the Child (CRC) reinforce this responsibility by challenging the States' Parties to uphold the child's right live in a safe, clean and healthy environment and to engage in free play, leisure and recreation in the environment. According to the CRC, a child's well-being and quality of life is the ultimate indicator of a healthy environment, good governance and sustainable development (UNCHS, 1996; UNICEF, 1992; UNICEF, 1996, 1997). If the goals of sustainability are not achieved then it will affect children more profoundly than other members of the global community. There is clearly a convergence and in many ways a symbiotic relationship between the principles of sustainable development and children's rights.

The next section of the paper will present four vignettes of children living firstly in PNG and then the Cook islands.

Site One: Papua New Guinea

The world's second largest island, Papua New Guinea (PNG) is still known today as one of the few wild places left on earth. With its impenetrable forests and remote steep mountains it has managed to fend off the trail of many colonists who wanted to extract its great wealth.

New Guinea has several climates, ranging from the tropical humidity of the coastal lowlands, and the temperate cool of the mid-mountains zone, to the Alpine chill of the summits. There are over 700 distinct languages in PNG amongst a population of 4.6 million; many languages are still undocumented. Since Independence from the Australian government in 1975, PNG has endeavoured to build a modern western democracy. Unfortunately, due often to government mismanagement and the ongoing battle with trying to maintain infrastructure in remote regions, many New Guineans have had to return to more subsistence rural lifestyles as the only means of survival. New Guineans are the customary landowners of their land and while unemployment is often quoted as being as high as 90 per cent - the majority of locals have the option of returning to their villages where food and housing is available. Poverty is becoming a way of life for many living in PNG and this trend is escalating. For many, the geographic isolation has seen the loss of minimal service; for others, social and political isolation has lead to serious corruption and crime. This is supported by O'Collins who recently wrote:

> Over the past two years Papua New Guinea has experienced an economic downturn, exacerbated by political instability, natural disasters, and a general deterioration in administrative capacity. These problems have seriously affected marginal households and communities already on the edge of poverty (2003: 3).

Infant mortality rates for children under five as taken from latest UNICEF data is stated at 112 per 1000 or 1:10. This is overwhelmingly high when compared to Australian children, 5 per 1000 or 1 : 200, who live only 2 hours flight away. Data concerning school enrolment during compulsory years of schooling (under 15) are also revealing, with 79 per cent of males and 67 per cent of female children enrolled in primary school. According to statistics, 43 per cent of the PNG population does not have access to piped water, 46 per cent does not have access to electricity, 91 per cent has no access to a telephone, and the average household size in PNG is eight - the highest in the East Asia and Pacific.

The two children represented in the paper come from two villages in two distinct provinces of PNG. The first child, Karo, lives in Hanuabada village, an urban village in the National Capital District that contains the capital city of PNG, Port Moresby. The second child, Betty, lives in Siremi village, a rural village in the

Oro province, the closest town being Popondetta (North-east of Port Moresby).

Hanuabada Village

The Motu Koitabu people of Hanuabada village are the traditional landowners of National Capital District (NCD), the city of Port Moresby. The coastal dwellers of this Indigenous group are the Motu (meaning island) and the land dwellers are known as the Koitabu. Motu Koitabu villages are spread throughout the NCD and the outskirts of Port Moresby. There are approximately 30,000 Motu Koitabu people living in the city.

The ever-expanding reaches of the city centre put constant pressure on the villages which, although in principle should have access to the resources available in the district (schools, health facilities etc.), find they are marginalised. A walk through the Motu Kiotabu urban villages reveals that most lack basic infrastructure such as adequate water supply, health services, education and sanitation. They have paid the greatest price for the benefit of the country - the loss of capacity to customary land that could serve to provide them with a livelihood.

Hanuabada village is the biggest of the urban villages. Around 15,000 people live in the predominantly Motu settlement. The village is divided into 26 clans - 18 Motu and 8 are Kiotabu. The majority of villagers live in sea houses built beside walkways. The accumulation of garbage on the shoreline, in the mangroves and under these houses built on stilts next to the piers, is one of the major problems facing the people of Hanuabada. It is not uncommon to see children swimming in a sea full of rubbish and human waste. When the tide goes out many of the houses are knee deep in raw sewage and mud - the smell is often unbearable, particularly in the wet season when the winds are often calm. The houses are clustered together so ventilation is a problem and there is a constant threat to the health of children who eat fish caught from the contaminated waters. Villagers complain of the change in colour of the sea, the result of bacterial build up from the daily deposits of raw sewage from the houses. Tidal flows, which have been constrained due to the development of road out to Elevala

Island, are not able to flush the mud flats. Consequently, the area is prone to infestations of malaria carrying mosquitoes.

Fig. 10.1 : Map of the Hanuabada Village by local youth.

Fig. 10.2 : Karo with her friends on the pier

Karo is 12 years old and in grade 5 at the local primary school. According to her uncle Karo's mother married a male (who was not Karo's father) in another village and moved away. For this reason she lives with her uncles family who has financial difficulties.

Her uncle believes she is just an average student at school - she tends to be very distracted, he said. Below is the picture Karo produced when asked to draw her village. She included in her drawing the sea, the canoes and the house where she lives. These she tells us these are important places for her.

Fig. 9.3 : Karo's drawing of the village.

Karo describes her drawing in the following way:

> *"The blue house is my house - I know all the other people in the other houses. The red house at the beginning of the pier is where my friend Maggi lives. She is younger and goes to school at Pari". Karo took us on a guided tour of her village during which time she shared with us her connection to particular places. Photographs were taken during our tour and her description emerged from our conversations with her during the time together.*
>
> *"My favourite places are the Church and Courthouse. The church because I go and pray there. It is Jesus's house - I want Jesus in my life. My other favourite place is the courthouse. I go there once a week. I go inside and listen to the talking. They are always fighting about the ground - this is my ground or this is our ground. I sit and listen and talk with my friends about who was right who was wrong. I walk to the courthouse by myself but there are always other children out the front who will go in with me. I also like to swim in the water near the pier, play sports, netball-*

seven girls sometimes 10 or even 12. Sometimes I play cricket with the boys who play on the road. Dangerous places are the gambling places - teenage boys play cards, drink beer. Also in town there are plenty of Rascals - older boys from other places who can hurt you. I am scared to go to town - sometimes I go with my parents, not by myself or with girls. I always stay in the village."

"Issues?- the sea is very dirty and around the piers there is a lot of rubbish. There are lots of plastic bags everywhere. The rubbish stinks especially when the water level drops. Sometimes they spray a chemical around that is to help with the mosquitos - but it smells and we are told not to play where they have sprayed. We don't have a toilet so everything goes into the sea. I hate this. I like to play with my friends in the water especially when it is hot but sometimes it makes my skin itchy. I wish we had a vegetable garden close by so I could grow stuff - but we have to walk a long way to where the land is and it is dangerous. There aren't any animals here - not even fish. I wish I could fish from the pier. It gets noisy at night - the drunk boys so sometimes I find it hard to go to sleep. The piers are very old and broken, they need to be fixed by no one has any money to fix it. Sometimes they get very slippery when it rains and if it is dark I am scared I will fall off. I like to watch television but we don't do that much. Sometimes the electricity gets broken and we can't even turn on the lights. Mostly I like it when it is quiet in the church. Sometimes I get sick. This year I had to lay down for a week. I had chicken pox last year I was very sick and didn't go to school for months. The kind of chores I do around the house include washing plates, carrying the water outside, putting things away, sweeping the floor, cooking rice (I do that everyday). In the future - I want to worship and pray. I like living in the village. Yes and I would like to learn to dance and visit other places. Also I want to be with my friends and not worry about the boys drinking and taking drugs."

Siremi Village

Siremi Village is located south of Popendetta in the Dobuduru grasslands. To get to the village you have to drive through ten-feet high elephant grass along a very windy track. At one point we are on along straight strip of the US airfield, at another we pass a number of mini Palm estates, the track then winds back into the high grasses. It takes approximately 30 mins by truck to reach the village from the main road. For children this would be at least a two-hour walk.

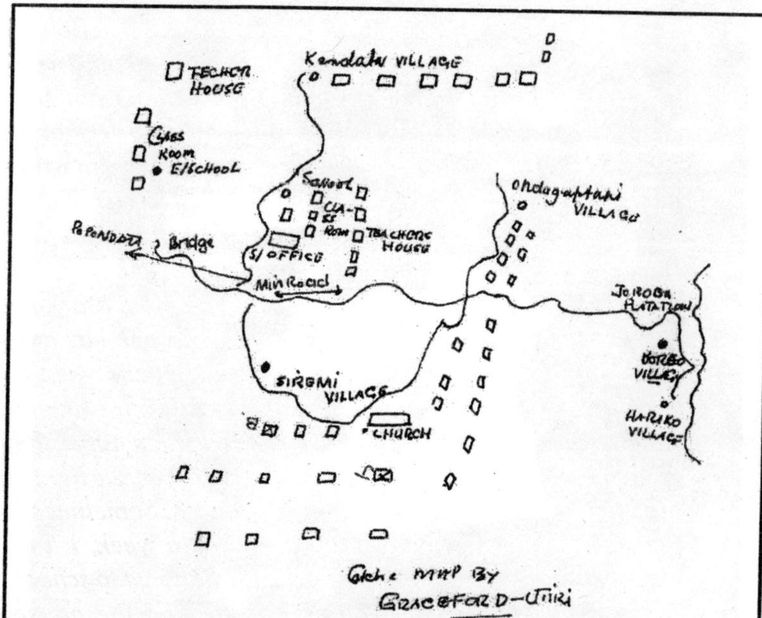

Fig. 9.4 : Map of Siremi Village by Mr. Utiri - Elementary teacher

Siremi village has 12 clans, 5 hamlets, 40 households and around 400 people. The majority of young adults work either on the mini palm estate or around the village itself. There are no motor vehicles in the village. To go to town (Popendetta) villagers must walk out to the main road and hope for a Public Motor Vehicle (PMV). PMV often break down on the 60 minutes round trip around the district and may not turn up. There are no medical facilities close by, a small medical outpost station is located 2 hours walk from the village, but does not regularly have staff. If a child

Fig. 9.5: Betty with her drawing.

or adult is injured or sick normally the villagers will take their chances on a PMV on the main road rather than go to the outpost. The only market for selling or buying produce, bank or pharmacy is located in Popendetta. Local villagers told us that to sell produce at the market was difficult due to the inconsistency of the PMVs – it was a long way to walk to the road with a barrow load of watermelon and then find you had to walk back with them again.

Access to fresh water is a significant issue for all villages on the grasslands with 93 per cent of the population's water source being the local creek or river. Only 6 per cent of the population have access to a well and only one village has a water tank. At the time of the study most villagers had no access to toilets and instead chose to use the bush or designated 'toilet areas' in the grasslands for relieving themselves. The lack of toilets and safe water has enhanced health hazards in the village with many water sources becoming contaminated. People in Siremi village complained of waterborne diseases such as 'sipoma' (skin disease), diarrhoea and in extreme cases cholera and malaria. With the large Palm Oil company moving towards leasing customary land for mini estates in the villages there is concern about the impact of pesticides and herbicides on these fragile water sources. Already in areas up stream from the small coastal villages contamination has lead to creeks and river being unusable and villagers having to travel long distances to collect water. The tension that exists between the sustainability of the palm oil industry, its impact on village life and its capacity to provide a disposable income for villagers is causing grave concern for locals.

Betty

Betty is 13 years old. She lives with her mother and father, 8 brothers and sisters and her grandfather. She is part of the clan Avetipa. She is in Elementary two at the school (equivalent to 2nd year of school) even though she has been at school for four years.

Fig. 9.6: Betty's drawing of the village.

"On my drawing I have drawn my house and the coconut palms that are around it. I drew my house on stilts because it is off the ground and we have to walk up the steps to get to it. I also drew flowers. I like the flowers in the garden. My favourite place is Port Moresby and the beach. I go to the beach with my friends or my dad on the weekends. We have a small canoe down at the beach (our relatives look after it) and we take it out and go fishing. I take my smaller sisters with me if I go. It takes about 30 minutes to get there. It is a nice walk. I like to play bowls, play with my sister and friends. My friends and I play ball games - like tiggy. If the ball hits you then you're out. Sometimes 10 or

20 childrens will play together in the village or on the school field. Sometimes we play hockey and volleyball at school. On Saturday sometimes we go looking for bushfruit and sometimes fishing with Dad at the ocean. Dangerous places - well the other villages and the roads - where there are noisy cars and trucks. My main job is to help my family is to fetch the water from the small creek near our house. In the future I plan to drive a truck. When I finish school I want to go to a big school like university and do some big job like accountant or something like that."

Site Two: Cook Islands

Rarotonga is the largest island in the Cook Islands with a land area of 67.2 square kilometres. The smallest island is Suvwarrow, at only 0.4 square kilometres with no inhabitants. Aitutaki is also one of the larger islands with a land area of 19.1 square kilometres. Unlike Rarotonga that is a high raised volcanic island, Aitutaki is an atoll made up of fifteen *motu* (islands) - the majority of local islanders live on the larger main island, the other fourteen motu are mainly small and uninhabited. It is a country of extreme differences with the highest peak of Rarotonga soaring 653 metres above the Pacific, and the six atolls making up the northern group being slivers of land so close to the ocean level that a rise of sea levels of two metres would inundate them.

The Cook Islands were first inhabited by seafaring Polynesians that came from both east and west. Tahitians settled the southern Cooks about 800 AD, while Tongans and Samoans settled the northern islands. From the Cook Islands, the people are thought to have migrated south to New Zealand, where they are also known as Maori. The country's first legislative council was elected in 1946, and independence was granted in 1965, though the islands remain self-governing in association with New Zealand.

The Cook Islands are one of only six countries in the world that have not officially ratified the Convention on the Rights of the Child (CRC). Basic UNICEF indicators of quality of life[1] for children and youth growing up in the Cook Islands illustrate that compared to other small Island nations provision is good. Unfortunately due to the lack of auditing data by the Cook Island government because it is not a ratified member of the CRC, UNICEF does not have information for all indicators. Population statistics reveal 8000 Cook Islanders are under the age of 18, with 2000 aged under 5 with the population growth across the country in the past ten years being 1.1 per cent. Infant mortality for under 5's has steadily been dropping over the past ten years (25 per cent reduction) with the current rate being 24 in everyone thousand children not surviving beyond five years old.

Over 50,000 tourists visit the Cook Islands annually, which is over twice the population of the country, and these visitors bring in approximately US $ 25,000,000 into the local economy. As well as encouraging employment into local tourist industries, it has boosted the livelihood of farmers and small business people and supported the development of social infrastructure such as medical centres, roads and banking. It is important to note that this tourism activity is predominantly focused on Rarotonga (and to a lesser extent Aitutaki); therefore much of its benefits do not seep out to the outer island communities.

Rarotonga

> *"My life here in the Cook Islands is perfect, everything fits in, it is a happy life. The food and fruit tastes so delicious. There are so many different things to do and the beaches are beautiful - so are the trees. The people are very kind and full of peace. All the children grow up peacefully. That is what I love about Rarotonga."* Ninia, age 10, female.

Rarotonga is the only true high island in the Cook Islands group. The extinct volcano is estimated to be 4.5 kilometres from ocean floor to summit. Rarotonga emerged from the sea 2 million years ago. Half a million years later the centre of the volcano collapsed, forming a caldera surrounded by peaks which became eroded into even more jagged shapes from weathering. The transportation of eroded material to the valley below has created a fertile ring of plains around the island. The island has three main physical zones: the coastal zone, a small flat area hugging the edge of the island where most of the housing and commercial activities are located; next are the slope forests that extend from 50 metres to around 400 metres above sea level; and then the cloud forest that occupies that area from 400 metres to the mountain summit.

Fig. 9.7 : Elizabeth's drawing of her island

Elizabeth

> *"My drawing is of the beach because it is my favourite place. I love to look under the water at all the sea creatures and make sure they are ok. We have to take care of the beach because the animals will die if we don't. In my picture you can see a coconut palm, beach and the sea and in the sea is rocks, fish, coral, seaweed and jellyfish. And at the front that's me. I am dancing at the beach. One of my favourite places is the Flame Tree - which is a restaurant. It is very good because people sing and dance and we get to eat really nice food. But best of all I like the beach. I like*

to go to the beach so I can swim in the beautiful blue sea. To me Rarotonga sounds like the drums beating loudly. It smells like a perfume you have never smelt before. It feels like paradise. It looks like heaven. The fruits taste delicious - like lasagne. I have a perfect life. I play fun games at home. There are not many places I don't like to go but I don't like the graves. They scare me. And the jail is bad especially when someone has to stay there I get scared for them. I hate having injections at the doctors and the hospital is very dirty. Oh yes, and the ocean I get scared when I go on the ocean. The places I play are the beach and the mountains. We have a taro patch in the mountain and I go there with my mum and we dig up the taro and I like to play in trees and smell the flowers. I think it is good that people come to our island. They need to learn our culture. But it is bad if they don't be nice or steal and stuff. We need to tell them not to pollute or leave rubbish. I think we should teach them about our beautiful island. The good thing about tourists is that they bring money and make jobs. The bad thing is that they leave rubbish and fight and get drunk. They should respect out culture and our language. When I grow up I would like to go to New Zealand, but I would come back."

Aitutaki

Aitutaki means 'Island of the Lagoon'. The island of Aitutaki is 250 kilometres north of Rarotonga. A triangular atoll a million years ago, Aitutaki's geology was changed by further volcanic events which formed the main island and two additional small *motu* in the lagoon. The main island is 12 kilometres long and boosts an array of native plants all tightly squeezed together. The central remnant of the volcanic activity is a small peak at one end of the main island that rises to only 124 metres. This bare hilltop provides excellent fertile terrain for supporting crops like bananas, pawpaw and arrowroot, and also a vantage point for viewing the breath of the turquoise blue lagoon and for this reason is a popular recreation spot for locals and visitors. Aitutakians have a tradition of story-telling and dancing and make and play the *pate* - the slit drum that accompanies most traditional Cook Islands dancing.

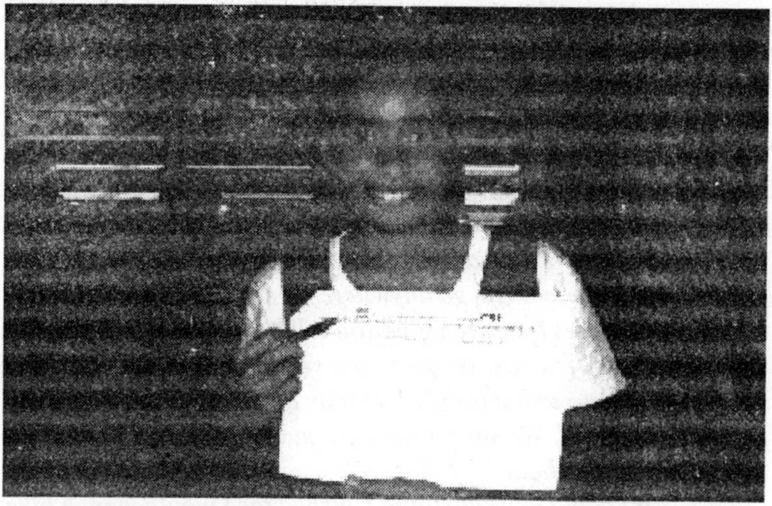

Fig. 9.8 : Teao filling out her survey

Teao

Teao is 11 years old and she is in Grade 6 at Aitutaki primary school.

> *"My drawing is of my island. Aitutaki is a large lagoon with lots of little islands. The lagoon has lots of fish and you can go fishing whenever you want. One-foot island is a famous motu on the lagoon. On my drawing you can see the trees and the hill and my school. My favourite places on the island are my school, my home and Akitua - the resort restaurant where we go sometimes. Oh and the hill. Sometimes we climb to the top and you can see the whole lagoon. I don't have any places on the island that I don't like to go to. I can go anywhere. Everybody knows you and sometimes they even give you a lift. But mostly I just walk with my friends. We go to the sea or the pool. We play at my house and if we want to play a game we come down to the school. We have a sand pit and lots of trees so you can sit in the shade."*
>
> *"If I could have anything I wished for I would want a bike and maybe some more shops. I would also like to have a computer. And I wish we had more television. I wish I could go and visit other places - I want to be chosen to dance. People come to the island to see it. I think its good that visitors come and watch us dance and they give us money. But they have to make sure they don't pollute the lagoon. Sometimes they leave rubbish and make noise. Sometimes they don't respect our culture. On the weekends we practice our dancing and we go to the beach for a swim. We like to watch the airplanes come in and see the visitors. Other times I like to go fishing or maybe collect shells."*

Consuming Spaces

The children of PNG and the Cook Islands are constructing and reconstructing identities through the hybridisation of local cultures while adapting to a changing global world. The stories of children in this paper are 'located' in a physical space tied to a fixed geographical locality, yet the relationships they form through space, time, culture and identity are deeply embedded within global political, economic, environmental and social constructs. The capacity for children to negotiate and participate in their environments is, I would argue, intricately tied to these temporal constructs and informs their experience of life at the beginning of the twenty-first century. While a specific discussion on space and gender won't be developed in this paper, gender is of significance to geographical constructions of space and place (Massey 1994: 180). That is to say, a predominantly patriarchal colonial past has influenced the gender relations and the structuring and experience of space, place and spatiality by the four girls in ways not addressed in this paper.

Returning to our metaphor of islands it is important to debunk the view that it is the impact of modern globalisation that has changed islands from being isolated and bounded social and spatial systems. In fact there are two very important issues to consider. Firstly, globalization is not a postmodern phenomenon. Networks of global social exchange, migration and colonisation have been an innate characteristic of human civilisations. This leads to the second issue, that is, 'Island' cultures have always been involved in extensive networks of regional communication and exchange with their neighbours (whether across oceans or hilltops). Their presumed isolation was an incorrect European assumption. The Pacific region as a case in hand has a rich history of migrating peoples. The Cook Islands for example was first settled about two thousand years ago with the fifteen islands being populated one by one by in a continuing process of ocean migration which had begun in South-east Asia around 4000 BC. Filtering east, travelling in double-hulled canoes which they sailed and navigated with consummate skill, the people who were to become the Polynesians spread out across the eastern Pacific.

Understanding culture and the influence of globalisation therefore needs to be constructed within a larger debate around the complex exchange of societies and the influence of those exchanges.

This is a critical issue if we consider the argument of critics of globalization who argue that globalization is just another form of cultural imperialism - that has at its aim the homogenisation of all 'local' culture to a universal western framework. In response to these issues, it is clear from the children's responses that they construct their identity through and within the multiple cultures and spaces they operate in, both local and global. For Betty who lives in a rural village in PNG seemingly disconnected technologically from the global world, her experience of globalisation is a daily tension between her family's need to survive in a degraded and hazardous environment, due in some regard to the impact of the multi-national palm oil industry, and the PNG government's lack of capacity to provide social infrastructure. For Teao, while living on her island paradise provides her with a great sense of belonging she is acutely aware of the world that exists outside of her immediate space - she is aware of the impact of tourists and tourism on her island, and like many of the children we spoke to on the island, she shares with us her desires to connect with the global community through technology and through her dancing.

Post-colonial criticisms pay host to the contestation of colonial political and social authorities who through struggles of power served to dislodge cultures of marginalised people. Both our research sites have a colonial history that bore witness to the divisive processes of an inscribing of a modern world order. For the children of PNG and Cook Islands the postcolonial hangover is most evident in the consequences of the colonial withdrawal on education, health environment and governance. Both Nations have been left to 'manage' their affairs with limited intellectual capacity, exhausted resources and patterns of migration and dislocation of peoples that have resulted in a largely offshore population. The impact of migration is particularly evident in the Cook Islands. More Cook Islanders live in Australia and New Zealand (around 40,000) than those who live on the Pacific Island nation. Elizabeth is keenly aware of the tensions of moving offshore (the 'brain drain' as it is often referred) and its consequences on her people when she notes she would like to go to New Zealand but that she *would come back*.

Virtually all environmental problems have underlying social, economic or political causes which either led to or can explain why problems have not been addressed (Satterthwaite *et al.*, 1996). Corruption, laundering of public money and lack of public accountability has thwarted PNG's effort for a smooth transition into an independent nation. The Cook Islands, due to its isolation, has found developing and funding its own infrastructure an enormous strain on a country which has limited capacity for economic viability - as a 'wedding destination' its economy clings gingerly to the whim of a unreliable tourist market. Consequently, both countries have limited capacity in terms of health workers (doctors, nurses) and even less potential to provide education to build capacity. The allure of the Cook Islands can be quickly tarnished after a short visit to the local hospital - as Elizabeth tells us it is dirty. In fact the advice given to all tourists is if they need medical attention they should leave the island immediately. The constant infestations of scabies, cholera, diarrhoea, and malaria in PNG and the extraordinary high instances of dengue fever, filariasis and diabetes in the Cook Islands are evidence of a need for governments to address environmental problems supporting these health hazards. The fact that most environmental problems that threaten the lives and health of children occur within low-income households implies that it is problems of poverty and inequality - limited household capacity to support basic services and education - that are the greatest threats. At a national level, the capacity for governments to provide those basic services that greatly reduce environmental hazards, or their health impacts, depends much on the level of prosperity of the national economy. In general, the higher a country's per capita income, the higher the proportion of a nation's population with access to piped water, provision for sanitation and health care (Satterthwaite *et al.*, 1996). The consequence is a dependency relationship by the fledging postcolonial nation with their Imperial 'other' and an over reliance on fragile economic (aid) arrangements. Children are acutely susceptible to these vulnerable environmental conditions: Karo who misses school for long periods due to illness and therefore, undermines her capacity to an educated future; Betty who spends up to 6 hours in her day fetching water from a contaminated creek a kilometre away from her village; Elizabeth and Teao who would need to endure a 6 – 8 hour flight (and great expense to their parents) to have access to a safe and well equipped hospital.

When we discuss the environmental needs of children in these and other post-colonial nations we need to be acutely aware of the importance of children's rights. Indeed the environment that 'holds' these children through their childhood should be a place that promotes their health and well being and allows them access to safe and secure environments where they can explore their sense of connectedness to their place and their participation in the global community; where they can play and participate in cultural activities; where they can associate freely with others without fear or threat of violence and where they have access to information, education and an opportunity to voice their aspirations, dreams and concerns (Chawla, 2002).

Nightmare or Paradise?

As the voyeur wanders aimlessly along the white sandy beach, the long blue lagoon stretches out before them, pineapple cocktail in hand, a sarong tied loosely across their waist, an admiring eye glances over to the local children who play seemingly impervious to their presence - the tourist is, through processes of consumption, inscribing and transforming this place as a space produced for them. Cleansed of the political it lays before them as a paradise to indulge and be seduced within. Theirs is a visual consumption - a space to be for a 'time'- the reality of the children's lives lies beyond this tourist gaze. Children's experiences of space in the Pacific Islands is layered over time, it is historical, complex and multiple - a hybrid of local and global cultures, a vast interlocking and non-interlocking set of networks and relations at every scale from local to the global. They construct 'self' through the daily exchanges within their local communities and the global exchanges that come to them through social, economic and political means. To understand how they engage in this use of space we need to conceptualise space as constructed out of these interrelations, as the simultaneous coexistence of social interrelations and interactions at all spatial scales (from the local to the global) (Massey, 1994). The physical environment is their place, their home - it affords for them the capacity to make sense of their world, and yet it also constrains and imposes challenges that have enormous impacts on their future lives. The children struggle between the comfort and yearning of the local and the pull of the global. A yearning, as bell hooks (1991: 147) aptly reminds us that we all have for a place we call home:

> ...home is no longer just one place. It is locations. Home is that place which enables and promotes varied and ever-changing perspectives, a place where one discovers new ways of seeing reality, frontiers of difference. One confronts and accepts dispersal and fragmentation as part of the constructions of a new world order that reveals more fully where we are, who we can become....

For the children of the Pacific, it is not a place to call home they 'yearn'. They are 'located' in 'place' in a very real daily way. What they talk of is the yearning to understand how they can come to occupy multiple spaces through simultaneous times within that place - moving from the local to the global - from the tribal to the post modern and back again.

NOTE

1. This data is based on the year 2000 statistics obtained by UNICEF (2002) in conjunction with the Cook Islands government and are published in the *UNICEF State of the World's Children 2002* report.

REFERENCES

Anzaldua, G. 1999a. *Borderlands = La frontera.* (2nd ed.) San Francisco: Aunt Lute Books.

Bhabha, H. 1994. *The Location of Culture.* New York: Routledge.

Beck, U. 1992. *Risk Society: Towards a New Modernity.* Sage Publications: London.

Chawla, L. 2002. *Growing Up In an Urbanising World.* London: UNESCO/ Earthscan.

Christensen, P., James, A., and Jenks, C. 2000. "Home and movement: Children constructing 'family time' " (pp.139-155) in S. Holloway, and G. Valentine (eds.), *Children's Geographies: Playing, Living, Learning.* London: Routledge.

Driskell, D. 2001. *Creating Better Cities with Children and Youth.* London/Paris: UNESCO/Earthscan.

Hall, S. 1996. "Who needs "identity"?" (pp.1-17) in S. Hall and P. Dugay (eds.), *Questions of cultural identity.* London: Sage.

Hooks, B. 1991. *Yearning: Race, Gender, and Cultural Politics.* London: Turnaround.

Luke, C. and Luke, A. 1999. "Theorising interracial families and hybrid identity: An Australian perspective". *Educational Theory,* 49: 223-250.

Massey, D. 1994. *Space, Place and Gender,* Great Britain: Polity Press.

O'Collins, M. 2003. *Isolation, vulnerability and governance: reflections on poverty assessments in Papua New Guinea, Fiji and Vanuatu.* Occasional paper Australian National University, Canberra. www.anu.edu.au/melanasia/ maev.html (accessed 22/05/2003)

Rapport, N and Dawson, A. 1998. (eds.) *Migrants and Identity.* Oxford: Berg Press.

Smith, E. 1998. *The Cook Islands.* Ewan Smith Island Image, Cook Islands.

Satterthwaite, D., Hart, R., Levy, C., Mitlin, D., Ross, D., Smit, J., and Stephens, C. 1996. *The Environment for Children: Understanding and acting on the environmental hazards that threaten children and their parents.* London: UNICEF/Earthscan.

Tomlinson, J. 1999. *Globalisation and Culture.* Cambridge, UK: Polity Press.

UNCHS. 1996. *An Urbanizing world: Global Report on Human Settlements 1996.* Oxford: Oxford University Press.

United Nations. 1992. *Agenda 21: The Rio Declaration and Statement of Forest Principles.* New York: United Nations Publications.

UNICEF. 1992. *Convention on the Rights of the Child.* New York: UNICEF.

UNICEF. 1996. *Towards Child-Friendly Cities.* New York: UNICEF.

UNICEF. 1997. *Children's Rights and Habitat: Working Towards Child-Friendly Cities.* New York: UNICEF.

Urry, J.1995. *Consuming Places,* London: Routledge.

Chapter 10

CYBERSPACE

A Contemporary Social Space

Chris Pawson, Eike Adams and *Ralph Tanner*

Social scientists often consider children's optimal development in Western societies to be dominated by the ideology of a broad and dependable social interactive forum. This is a predictable and sustained ideology derived from the impressive body of literature from diverse academic disciplines demonstrating the host of positive outcomes associated with positive social interaction (e.g. Hartup, 1996; Werner, 1982). However, the authors' varied experience of fieldwork in rural and urban communities in Western Europe reveals that the child inhabitants of many Western communities are increasingly subject to a wider range of social restrictions than previously experienced. Societal circumstances, such as increasingly perceived threats to children's safety and decreasing spatial and temporal opportunity, create a range of statutes preventing Western children from engaging in social interaction that are often far beyond the restrictions on the social interactive liberties which are part of traditional living.

Socially, children are at the bottom of the hierarchy in the majority of social space they enter or inhabit. However, any hindrance to social development created by this inferior status and disparity of social power can be ameliorated by the child's experience of peer relations, both at school and in their neighbourhood. Peer relationships are often a source of liberation and the social scientific literature demonstrates a wealth of evidence for their social developmental influence and role in later success (Bukowski and Hoza, 1989; Ladd, 1990; Parker & Gottman, 1989). However, it has been suggested that Western children since the

Second World War have gradually seen the erosion of these social networks and relationships that are so valuable to their development (Coleman, 1987). It is the authors' contention that unfortunately this decline in social interactive opportunity has continued unabated and even accelerated more recently in Western Europe due to two key contemporary developments: (1) the institutionalisation of education as a process of government-led rigid curricula and excessive assessment has resulted in increasingly structured daily activity within schools at the expense of valuable peer interaction in the classroom and school playground; and (2) a gradual decline in community activities and physical space coupled with an increase in the real and perceived risks to children in the space between the supervision of school and home further limits peer interaction.

The terrors that lurk outside the school gates or the home may not be any more hostile than those of Western children 20 years ago, nor of children in traditional societies. However, for the care-givers of inner-city Western children there is much perceived latent and unknown hostility when they are not being supervised, and these risks are proving enough to curb the traditional peer forum and the developmentally salient interaction widely recognised as so valuable. The reports of London teachers and parents demonstrate clearly that inner-city children in London are functioning within a social world in which time and space are increasingly regulated. To use Coleman's (1987) phrase, children are losing their 'social capital' in terms of social interactive time and space. A similar profile is evident from the research of the authors in other cities across Europe. In response to these data, the authors propose that the potential for developmentally salient social interaction with peers will continue to decline and will eventually become recognised as detrimental. Furthermore, the existence of children in traditional societies may be recognised, at least in terms of wealth of peer experience, to be considerably more advantageous in social developmental terms.

'Cyberspace': An Alternative Physical and Social-Interactive Space

"Across the world there is a passionate love affair between

children and computers... they seem to know that in a deep way it already belongs to them. They know they can master it more easily and more naturally than their parents."

Seymour Papert, The Connected Family, 1996

Despite this apparent decline in the spatial and temporal opportunities for social interaction amongst many urban European children, through the internet there exists an alternative 'space' in which they are being afforded more opportunity - 'Cyberspace'. Internet access via computers creates a space in to which children consciously or sub-consciously are entering (Suler, 1998). The parlance of internet activity has become one of physical movement and physical domains, adopting phrases such as 'going' to 'sites' or 'entering chatrooms'. The internet, or 'cyberspace' is a considerably larger, more diverse, and for many, more navigable, space than almost any physical space experienced by inner city Western children. Not only does cyberspace provide a space within which to explore outside their proximal physical world, it can also create a social interactive forum well beyond that of the child's proximal social world.

Exploration of the physical world via information on websites provides many developmentally significant opportunities and experiences for children. However, the focus of the current authors' research is the child's use of cyberspace as an extension of their social world. More specifically, the unique characteristics of internet communication provide a multitude of social worlds and communication tools in cyberspace such as chatrooms, email and instant messaging, all of which provide a plethora of social opportunity.

The Proliferation and Popularity of Internet use among Children

Children from a very early age have access to web-based information and social networks through computers at home and school, and whatever the intentions of adults where these computers are located, it is in practice impossible to control what, and to whom, they have access. It is becoming, if not already has become, an almost universal childhood skill which leads to a space within which children are empowered with reversal of the traditional parent-child relationship as the internet-savvy child takes on the role of cyberspace guide to the parent.

The proliferation of access to computers in the schools, homes, and even bedrooms of children in the Western world, has provided children with a very popular and unique space within which to operate and socially interact. To illustrate, when a nationally representative sample of American children aged 8 to 13 were asked which medium they would choose to bring with them to a desert isle, more chose a computer with Internet access than any other medium, including television (Rideout, Foehr & Roberts, 1999). Furthermore, a recent British survey conducted by the London School of Economics reported that 75 per cent of 9-19 year olds have accessed the internet from home, 92 per cent from school and 72 per cent use it to communicate with others (Livingston & Bober, 2004).

The Social World of Cyberspace

Advances in the utility of the Internet, beyond the shared provision of information it was originally intended for, have significantly diversified the nature of relationships and social exchanges within cyberspace. Studies of email use among adults have consistently demonstrated the interpersonal side of cyberspace utility. Adult users commonly report that they socialize, maintain relationships, play games, and receive emotional support with people they know via email (e.g., Finholt & Sproull, 1990; McCormick & McCormick, 1992; Rice & Love, 1987). There is also evidence that social interaction occurs amongst individuals who have never met face-to-face (e.g., Brennan, Moore, & Smyth, 1992; Myers, 1987; Wilkins, 1991). The accounts of users also make it clear that the relationships experienced in cyberspace are genuine personal relationships in the eyes of the participants (Parks, 1996). Although the aforementioned data relate specifically to adults, there is also some limited evidence to suggest that children do utilise and perceive the social world of cyberspace similarly (Gable, Gross and Juvonen, 2002).

Beyond email, children are now able to synchronously interact with known others through instant messaging (IMing) whereby they are informed when friends are online and can then chat to them through text boxes on screen. This allows for intimate dyadic or peer-group communication in real-time with privacy. This channel of cyberspace interaction was recently found to be the primary medium for contacting friends amongst 20 per cent of American teenage internet users (Gable, Gross and Juvonen, 2002). Gable *et al.'s* study also revealed that the majority of their participants indicated they were communicating with IM partners "everyday" or "almost everyday" and that the median interaction through IMing was 30 minutes. Children's motives for using IMing convey the social and personal nature of participants' interactions and lend support to the authors' hypothesis that cyberspace provides a valuable alternative and even compensatory social space for children. The most widely reported motive for IMing was "to hang out with a friend," endorsed by 92 per cent of instant messagers interviewed, and 74 per cent of instant messagers reported IMing because they were bored (Gable, Gross and Juvonen, 2002). According to intimacy theory (Reis and Shaver, 1988), such interactions should be positively related to psychological and physical well-being.

Although the vast majority of IM interactive partners are known to the child, Gable *et al.* also found that 12 per cent of IM partners were first encountered in cyberspace. Therefore, not only does cyberspace provide an extra forum for peer interaction amongst children who may share other social space through IM, it also provides the sole forum for a number of peer relationships. It is this potential for allowing children to initiate peer contact amongst unknown children that is perhaps the most interesting and unique aspect of cyberspace. Beyond the children for whom IMing provides introductions to new interactive partners, many more are able to utilise cyberspace as an opportunity to participate or merely observe social interaction through 'chatrooms'. This option provides twenty-four hour access to an extended network of peers that would frequently be impossible to meet face-to-face due to geographical distance. Chatroom users are typically able to enter at any point and review the current conversation before typing their contribution and submitting it for the rest of the chatroom community. See Fig. 10. 1 for a typical chatroom layout.

Fig 10.1: Example of child chatroom layout

Social interaction within cyberspace appears to be similar in several ways to traditional means of youth social interaction (Brown, 1999). Furthermore, on-line communication is reported to be mainly devoted to ordinary yet intimate topics (e.g., friends, gossip) and motivated by a desire for companionship (Gable, Gross and Juvonen, 2002). However, Cyberspatial interaction clearly differs from interpersonal interaction in the 'real-world'. It can be argued that the social distance between individuals or groups in cyberspace is almost entirely unique to this environment and therefore, introduces unique concepts of how and when humans can interact. In so doing this environment creates phenomena and raises new opportunities for the way individuals relate to one another (Lea and Spears, 1995). It is these phenomena and opportunities that the current study aimed to explore.

Unique Characteristics and Limitations of Social Cyberspace

If the human senses can be seen as having developed a specific utility in interpersonal relationships, the use of cyberspace requires no more than the use of two fingers and visual ability. It is quite outside the use of touch, hearing, smell, speech and personal appearance which influence so many human relationships. Furthermore, it is perhaps difficult to conceptualise the deployment of social interactive prerequisites such as prosocial behaviour, reciprocity or empathy whilst sitting in front of a computer screen. Intuitively this suggests that no social skills at all are used and that a computer can offer no potential for the observation, practice or creation of social interaction. However, beyond the screen, keyboard, and mouse, is a space within which children's social interaction is liberated from the confines of physical locality and they can employ all of these aforementioned complex social behaviours regularly. Indeed the authors argue that these behaviours occur overtly in cyberspace with at least the same regularity and complexity observable among peers in a school playground.

Research has likened cyberspace to a kind of virtual laboratory for exploring and experimenting with different versions of the self and opportunities for the projection of desires, frustrations, anxieties, and often the fantasies they might fuel (Suler, 1999). In contradistinction to all their social experience as children, they obtain through the use of cyber space equality, free expression, status enhancing exchanges and freedom from supervision. Nothing is considered except what they type in and provide access for; they are in total personal control of what they present and to whom. Each of these liberties is due to the anonymity and social distance of all of the potential interactive partners in a chatroom (Bargh, Fitzsimmons and McKenna, 2002). Each participant is free to experiment with constructions of the self and then disengage at anytime without fear of social retribution and join another social group almost immediately with very little potential for judgment based on prior behaviour. This unique opportunity may provide a similar role to games and other forms of play which allow children to develop social skills for later life but without the costs and potential dangers of making mistakes.

Indeed research concerning social use of cyberspace demonstrates a variety of socially supportive messages and positive outcomes for children. Cyberspace can provide children with special needs with valuable opportunities to network with other children and share their feelings about having a disability without fear of being stigmatized (Hasselbring and Williams-Glaser, 2000). Likewise, data from adolescent girls indicates that sharing information in cyberspace, including sexual experiences, may actually encourage greater caution and patience when making sexual choices in real life (Brown, 1999).

Cyberspace lacks traditional aural and visual cues such as vocal qualities and bodily movement and has been posited as lacking the informational richness of face-to-face communication (Kiesler, Siegal and McGuire, 1984). However, the absence of these social cues does not necessarily limit social interaction through restricting communication of relational and personal information as might be predicted by social presence theory (Rice, 1987; Rice and Love, 1987; Short, Williams and Christie, 1976). Instead, cyberspatial interactions employ innovative social cues and may even provide greater opportunities for deeper, more liberated, and relaxed social relationships than face-to-face environments (Bruckman, 1992; Sprouil and Kiesler, 1991; Wilkins, 1991). Cyberspatial interactions adopt a range of text-based methods to express emotional and physical information. For example, users employ 'emoticons' or 'smileys' through the use of text characters to indicate an emotional signal or reaction, e.g. " :-) " to indicate a smile, or, " ;-) " to indicate a smile with a wink. In addition, users frequently express emotion and metacommunicative intent by embedding abbreviations in text and may type in capitals rather than lower case to indicate raised volume, e.g. " LOL " = Laughs out Loud. The use of embedded phrases are also widely employed as if stage directions in a script to indicate a physical reaction, e.g. "this is so frustrating *clenches fists and teeth*". This use of language is often demonised, which Merchant (2002) attributes to the 'narrowly-defined version of literacy in the school system'. However, more possible readings of these linguistic features are possible if one draws attention to the creativity and innovative character which can be associated with this evolving language (Merchant, 2002).

In summary, the benefits of peer interaction such as feeling understood and appreciated, intimacy, and shared connection with others are especially strong predictors of well-being (Reis, Sheldon, Gable, Roscoe and Ryan, 2000). There is considerable evidence to suggest that the developmentally influential peer interaction and social capital of Western urban children has seen gradual decline in recent years. This is largely due to limitations on temporal and physical space at school and in their neighbourhoods. In light of research highlighting the developmental importance of peer interaction, it seems likely that such decline is likely to be reflected eventually in the cognitive, physical and social development of these children. However, the authors argue that despite limitations, social interaction through cyberspace may provide an alternative compensatory source of developmental opportunity. The purpose of the current study was to examine to what extent cyberspatial social interaction might mirror or diverge from, that of physical space, thus providing a potentially beneficial alternative or additional forum of social interaction. Specifically, the study examined the processes of initiation and maintenance of children's social interaction and relationships in cyberspace.

Method

In examining the potential for considering cyberspace as an alternative social space for children researchers were interested in the similarities and differences between children's social interaction in cyberspace and their lived environment, especially in relation to the formation and maintenance of peer relationships. To this end, a thematic discourse analysis of 280 chatroom conversations was conducted. To ensure a grounding of the analysis in the data, the authors followed methodological guidelines in the process of their analysis which were theorised by Strauss and Corbin (1990).

The data used in the current study consisted of downloaded conversation threads occurring in children's chatrooms over a two-week period. Both chatrooms were managed and administrated in the United Kingdom and stipulated that they were only for use by children under the age of 16 years. The authors selected two chatrooms which provided records of all cyberspace interaction and published the chat on open access websites so as to avoid the researchers needing to actively 'sign-in' or participate in cyberspace in anyway.

Biographical information was not available for all participants in the chatrooms, however, the available data revealed a diverse population interacting within cyberspace. Children ranged in age from 6-16 years and entered cyberspace from locations in Western Europe, Hong Kong, Australia, and across the United States of America. Both sites required children to sign-in to the chatrooms and create a username. Data revealed that the duration of children's membership of the sites ranged from 1 to 546 days. The frequency of cyberspace communication varied from an average of 1 cyberspatial communication a day to an average of 80 per day. Similarly, total communication ranged from 1 in total for new members to in excess of 19,000 by the older members.

Few studies have addressed the ethical issues surrounding the use of on-line conversations as data, however, the authors consulted with, and abided by, the guidelines of the British Psychological Society in the collection and analysis of this data. By posting messages in a public forum, most participants in on-line communities realize that anyone can read their messages, and many people will communicate outside the chatroom with the other participants if they wish to have a more private conversation. Studies should be sensitive to ethical issues involved in collecting data in this manner, particularly with children and despite the public nature of published online discourse, the authors have deleted any details of participants in respect for their privacy, including names, sign-in names and specific details of individuals' location. It should be noted that in printing excerpts of chatroom communication, the authors have left spelling and idiosyncratic grammar intact to give a flavour of the language utilised in cyberspace.

Discussion

There are three aspects in the chatroom talk which the authors focused on in examining children's creation of a social space online: (1) what they talk about, and how this talk might be different from what discourses are possible in their 'real world', (2) how they talk about issues, and simultaneously initiate and maintain their interactions, alliances and relationships in cyberspace, and (3) to what extent the social interaction in cyberspace may fulfil some of the developmentally important functions of children's peer interaction in 'real space' (Asher and Parker, 1989), such as:

- Fostering the growth of social competence
- Serving as a source of ego support
- Providing emotional security in novel or potentially threatening situations
- Serving as sources of intimacy and affection
- Providing guidance and assistance
- Providing companionship and stimulation

Initiation of Social Interaction in Cyberspace

From the frequency with which most users visit the chatroom it becomes apparent how important this interaction seems to be for these children. Obviously the number of posts each user has posted is relative to the time in total which they have spent in the chatroom, but it is impressive to see that most users have a number of posts in the hundreds or thousands, with some users having close to 20,000 posts. This illustrates that the chatroom is not only an infrequent pastime, but is a significant part of some children's social interaction.

In many ways, the formation of relationships in cyberspace is similar to the way in which these relationships are formed in the lived space. They follow similar phases, and require investment not only at the initiation stage, but also to maintain the relationships. Initiation and maintenance of social contacts in cyberspace are managed with the help of certain strategies, such as welcome notes and information provision for new users in the chatroom to initiate first contacts, or a hierarchical categorisation of relationships into general social contacts and friendships.

Welcome Messages

When a new user appears in the chatroom, one of the more established users will often welcome them. The new user is taken into the group by this gesture, implying a very open social structure of the group which seems generally supportive. Being welcomed by a more established member does however also fulfil other functions; for instance it indicates a hierarchy within the boards, of older members establishing themselves in a function which could almost be described as an internal mediator or host (in some ways similar to the webmaster, who has this function externally, without taking part in the discussions). As will be discussed, however, this welcome is not the same for all new users, but heavily reliant on their presentation of themselves to the group.

A typical introduction from a new user to the group would involve some disclosure of personal information, as seen in the Extract 1 below, potentially followed by a request for replies, as in Extract 2 ("Please write back!"):

Extract 1

> AC (1st post): *"Hi i'm from Germany. Can some People speak german in this chat? i can speak a little bit English. I not so good i have a Mark 2 in English. But this yeahr I have a trip in London, I learn English there too! Please not angry, i can't speak/write English. But i'am learning. see you later!"*
> AD: *"i speak rough german, i started in September, love to hear from you."*
> AE (9th post): *"gutentag!!"*
> AF (19th post): *"Gutentag!!! (i only no a little of german!!) hello welcome to the message boards. Have a lovely time*

in England. Nice chatting to you!!"
AG(967th post): *"Sehr gut! haha... you're very good at English. I can speak German quite well."*
AH (1st post): *"Hi, I'm S. I can speak a little bit german. I think no is – oh i can't remember! I used to know it. Can you tell me how to say this in german. If you do, i can say it 2 my brother, and he won't understand! Rightr, here is the sentence - 'Go away you annoying boy!' thanx so much!"*
AB(12764th post): *"Guten Tag! :D Er...that's all I know : S Your English is very good!"*

After this exchange, the topic changes, and the new user doesn't seem to appear again either, but the extract illustrates nicely how she is welcomed into the new group by means of a polite and friendly exchange. Another user expresses a welcome, and two of the well-established users immediately compliment the new user.

In the second extract, someone is involved in the group readily via an interesting discussion about drug tests in schools that he starts off:

Extract 2

AL (2nd post): *"My school don't have drug tests but i know for a fact that some people have them. Please write back!"*
AZ (1662nd post): *"Cool screen name"*
AX (21st post): *"I don't see the point of the drugz tests, what next? Body guard for teacher's?"*
AY (87th post): *"My school is the same. I know loads of people who have them, but we don't have drug tests. I was actually quite shocked when those school girls were arrested for having them."*
AM (1199th post): *"There isn't drug testing in my school either, but I understand why it's necessary; after all, it's a health issue at the end of the day, and whereas I would HATE it if I had to have a drugs test, because I would feel so guilty even though I had no reason to, I would also*

> *appreciate why some people are against them, because if you've never taken drugs before, it's natural you're going to think the testing for all is unjustified!!*
> *Luv, AM xxx*
> *PS: Welcome to the boards!!! Have a fantastic time and enjoy yourself!!! I LOVE your screename (sic)!!! LOL!!! :) :) :) xxx"*

In this extract, again the new user is welcomed and complimented by the two most established users, and interestingly AZ doesn't actually even take part in the discussion, her only contribution consists of complimenting AL on the screenname. This, and the fact that it is only established users (with at least 100 posts) who express a welcome in some shape or form, might indicate that some of the function of these welcome notes lie in the consolidation of an established hierarchy according to number of posts or length of time spent in the chatroom.

On other occasions, new users sometimes introduce themselves to the group in ways which are slightly unusual (compared with face-to-face interaction) in that they don't offer any information about themselves personally at all, but at the same time ask for an intimate relationship in return:

Extract 3

> VA: *"who want's to be my friend?"*
> VR: *"Hi who wants to be m8s?"*

Rather than just asking for contacts or a chat, they actually ask for friends, but without giving any idea who they are. From the new user's point of view it can be argued that this chat room is seen as a social space which might provide an opportunity for an expansion of their own social circle. However, it seems a little unorthodox to expect someone to immediately engage in 'friend-ship' if they have no information about the potential friend whatsoever, even at this young age, where friendships are much more easily formed than in later life. And, consequently, these postings were just ignored. It is possible that this points to the chatroom as a social space being similar to the lived social space. It is relatively unlikely that a child would approach a group of unknown peers with the line "who wants to be my friend"; if they did, it is likely that this situation would end in disregard at least, if not ridicule or worse. Extract 4 demonstrates the value of a brief introduction through the use of what cyberspace users term 'a/s/l', short for age/sex/location. Through volunteering the smallest personal detail via her a/s/l the user in Extract 4, a 9 year old female from the U.K., is greeted considerably more favourably than VA or VR in the previous extract. The a/s/l's of the respondents RB and RC also clearly illustrates the diverse location of cyberspace users, coming as they both do, from the USA.

Extract 4

> RA: "*hi im (uses real name) im 9 years old and im lookin for a freind so please reply to me !!!!!!!!! love (real name) xxxx 9/f/UK*"
> RB: "*hey I'm (uses real name)! I'm almost 12 years old I'll be you friend! 11/f/usa*"
> RC: "*hey whats up i would love to be your friend how are you doing?? 13/f/cali*"
> RA: "*hi (RB) do u wana be my freind????9/f/UK*"
> RA: *Hi (RC) wana t6alk then?????????9/f/UK*

The chatroom does not then provide a free-for-all social environment, but is guided by rules of social interaction which will be similar to other social spaces. Rules for the initiation and formation of peer relations might be very similar to those in real lives, although ultimately maybe more flexible: teenagers in real-life situations are often 'stuck' with a particular group of peers, e.g. due to schooling. In the chatrooms they can choose the chatroom or the particular conversation thread by topic, hence enhancing their chances of finding like-minded peers.

In general, new members on the message boards are made very welcome, which can be illustrated by postings whose purpose is to make 'life on the boards' easier for newcomers, and to disclose linguistic particularities of the chatroom. An example of this is provided in Extract 5 whereby an established user offers a guide to the meaning of emoticons and abbreviations, in order to ease the entry to the discussion and not be exclusive. The use of these specific linguistic features fulfils several functions. Not only are they used out of economical time management, but can also be tools of what Merchant (2002) termed the 'colonisation of new territories and mapping of contours of a new communication landscape'. Children in the chatrooms consistently demonstrated their reliance on emotions to overcome the absence of non-verbal behaviour and emotional expression that is so central to face-to-face communication.

Extract 5

"Here are some tips!
:) = hAPPY smiley *:(= Sad Smiley* *:D = Even hapier smiley*
;) = Winking Smiley *>:(= Angry/Cross smiley :P = Sticking tounge out smiley*
:| = Normal Smiley *:o or :O = amazed Smiley:S = Confussed Smiley*

Newbie (You) = New Person Oldie = 3-6 months
Bronze Oldie = 6-9 months Silver Oldie = 9-12 *Golden Oldie = 12 months+*

---

u= You *r= are* *y= why*
(uses real name) xx"

Beyond initiation of interaction within cyberspace, the authors observed two main mechanisms at work to initiate, maintain and consolidate more formal relationships: the expression of solidarity in situations of conflict, and the overt declaration of a particular relationship or friendship, being 'e-buddies'.

E-buddies

An explicit way of forming alliances and friendships is apparent on the boards in the shape of by naming someone their 'e-buddie', meaning a (particularly close) friend on the boards. By seeking out these friendships, complex social relationships are formed, and support and encouragement as well as reciprocal affection are provided.

On one of the chatroom sites the authors came across a relatively long thread in which the users discuss their 'bestest e-buddie', sparked by one girl who wants everyone else to nominate one person as their 'e-buddie'. Many users oblige, and it is interesting to see how relationships get consolidated in this exchange, but also how precarious some of the relationships are. It seems useful for the discussion to provide the extract in its entirety at this point:

Extract 6

> AN (3262nd post): *"Who is your BESTEST BESTEST E-buddie? You can only chose one...*
> *Mine's AB. We have loads in common too...*We're both Friends nuts.*
> **We both have black layered hair. *We both like Music. *We both have unusual Nick-names!! *We both love bananas!! They're dead healthy, ya kno~ plenty of vitamins and fibres.... :D"*
>
> AB (12779th post): *"Heyllooooo :P*
> **You got that right~could Friends BE any cooler? *We even look alike...? O.o *Of course :D Completely different types of music though...I think the only type of music w have in common is Xtina! *Hehe, yes *Well, YEEEAHHH!!!*
> *And apart from that, they are the best things ever!*
> *Love yoooo xxxx"*

AB is one of the most senior (according to number of total posts) and most popular members on the board. Hence AN's initiative might be a way of lifting her own status by means of association, as she wants to be nominated by AB as HER best e-buddie too, but isn't. In fact, AB responds, but possibly strategically makes no nomination herself :

> AN (3273rd post): *"Am I your bestest ever e-buddie too...? <prays>*
> (she does not get a reply to this post)
> AO (49th post): *"Mine is AP and the girl that is really into onions."*
> AS (3846th post): *"Me thinks that would be AR?!"*
> AP (19581st post): *"Thanks, hunni! :D"*
> AT (3251st post): *"L. as in AP*
> *i know her offsite, we have loads in common and i can tell her anything and i hope she feels the same way :)."*
> AP (19572nd post): *"Awwh N- so sweet! :D :D*
> *You know I cantell you anything. ^_^ Luv ya!"*
> *AT (3271st post): "yup, im very sweet :P surely you know that?!?!*
> *lol, hope so :P"*
> AU (1022nd post): *"AB"*
> AP (19572nd post): *"Ooh. I can't choose between 3...so I'll have to say: AW [says real name], AT [says real name] and AO (what's your real name again...?!) :) Although there's many people who come close. ^_^*
> *AW [says real name] and AT [says real name] are 2 of my best mates off the boards, too. ^_^ Luv ya, you two! :D :D*
> AW (696th post): *"aw AP dude! :D :D u'r making me get all bigheaded again :P ure my best mate on the boards and one of my best mates off the boards :D dudeness (H) got 2go finish getting ready for school now : (ttyl :D"*

For all the users, naming their best friend publicly on the board aids to consolidate relationships with each other, especially as in most cases this election will be reciprocal. Just as in their lived social environment, within a bigger social group smaller alliances and friendships are formed. One particularly interesting example is the exchange between AP and AT as they also know each other 'offline', in real life, and have formed a relationship in cyberspace as well as real space. Knowing each other in real space seems to contribute to a consolidation of their relationship in cyberspace, and probably also *vice versa,* as they are sharing two distinctly different social spaces which merge in their relationships. This maps onto one of the important developmental functions which children's friendships fulfil, according to Asher and Parker (1989) to serve as sources of ego support.

Finally, examining the initiation of interaction, Extract 7 below very clearly demonstrates many of the aforementioned factors of initiating conversation successfully, welcome messages, and the formalisation of relationships through becoming 'e-buddies'.

Extract 7

TA : *"I reckon we should be allowed to wear our own clothes to school every day!! By the way this is my first time on the board!!"*

TB : *"I don't as it would create rivalry between children due to the clothes they wore. Like, if someone got Total 90s and some1 else has better shoes, then they would b at war, the clothes children wore would also create gangs"*

TE : *"Welcome to the boards"*

TA : *"Thanks TE!!!"*

TF : *"I agree, school takes away individuality, and freedom to wear what you want..."*

TG : *"Hi!! Welcome to the boards, dude!!! LOL!!! Have a fabby time!!! Its great on here!!! I LOVE your screename!!! I know what you mean. It would be fun to wear your own clothes to school each day, but it'd be a little unfair I think, because you will always have people dressed upin designer gear, and people wearing less 'approved' stuff, so them people would have a hard time. That's just my opinion anyway!!! LOL!!! :):):)XXX*

p.s. : *Do you want to be e-buds? (E-buds are jjust like friends on the boards, just in case you didn't know, as you're new!!!) LOL!!!:):):)XXX XXX XXX XXX XXX XXX XXX*

*Luv, *REAL NAME*XXX*

TA : *OK den I'll be ur e-bud!!!!! Thanks 4 liking my screen name!!!*

TG : *Cool!!!We're e-buds then!!! HiYa dude!!! LOL!!!:):):)*
You're screename is well cool!!! I LOVE it!!! It's really fabby and original!!! LOL!!!:):):)XXX
See you around,
**REAL NAME* XXX*

A mere declaration of friendship, as is the case outside of cyberspace, is not sufficient to actually maintain social interaction, such that it may be beneficial in the terms proposed by the authors. Looking back at the functions of children's friendships provided in the introduction to this analysis, most of the functions Asher and Parker (1988) mention are linked to the provision of social support, either in terms of guidance and support, affection, emotional security and the provision of a secure alliance. In the next extracts, the nature of interaction in cyberspace via the provision of social support is therefore highlighted.

The Functional Utility of Cyberspace Relationships – Social Support

The authors' contention that cyberspace may serve to provide a valuable social forum was clearly supported by the consistent emergence of children's solicitation and provision of social support from other cyberspace users. The data demonstrated that whilst in some extreme cases cyberspace may provide the principal form of peer social interaction in the absence of peer relations in physical space (Extract 8), cyberspace was largely being used as a forum for additional support (Extract 9).

Extract 8

KA : *"i really wonna make friends and i dunno how to....i'm shy and i wonna hang out with ppl....every saturday and friday night i sit at home alone and watch tv! what should i do to have more fun in my life!!!!"*

KB : *"I'm sure that you go to school. You have classmates that might be as shy as you are. As long as you be in school you dont have anyfriends[classmates]. I'm sure someone would like to be your friend. I know its hard makin friends but if you really want friends then try hard to make some. I know what you mean about being shy cause i'm shy too. So thats one thing we both have in common lol."*

KC : *"hey im always lookin for more friends"*

KD : *"So am I."*

KE : *"ha thats funnie because i do the same thing to beacause i dont have friends either"*

In response to disclosure of an incident between one child and her friends from school, two children (NB and NC) respond:

Extract 9

NB : *my ex friends were like that last Monday my friend had a party and didn't invite me! I wasn't exactly bothered but they ignored me all day & at lunch my close m8s came up to me cos I sed the others have been ignorin me & my other friends didn't like it cos I "talked" to sum 1 else so they evideted me from their gang cos of it! Im not their m8s and I feel great cos im free to talk to ne1 else well I was b4 but it feels real gd!! I no that probablyt didn't help! But never mind im sure it will work out!!*
As Dori from finding Nemo says – Im glad I goit that off my chest!!

NC : *You have friends on these boards so sdon't worry! -xxx-*

Both Extracts 8 and 9 demonstrate the value of cyberspace in providing empathic responses, guidance and a cathartic forum in which children can voice and share their experiences. The shared experiences and guidance provided by KB and NB unmistakably illustrate these aspects of cyberspace utility. The reassurance provided by NC regarding the presence of 'friends' on the boards regardless of the tribulations with offline peers is also indicative of the perceived value of cyberspace relationships amongst these children and the validity of the authors' contention that they can be viewed as compensatory by some children.

Frequently the authors found that when soliciting support in the chatrooms involved issues with children's offline peers, significant levels of disclosure about complex dilemmas and contextual information occurred. It appears that even amongst children's close offline friends, social support frequently cannot be solicited from within the physical space, and relationships with other cyberspace users who are removed from the actual social situation can provide very valuable guidance and support. It is evident that there are many issues for which the anonymity of the cyberspace forum is ideally suited (see Extracts 10 and 11).

Extract 10

HA : *"Okay, so i have this friend, let's call her Joan, and I'm not sure if I should be her friend or not. I mean every other dayu she's mad at me for something, and sometimes what she says really hurts my feelings. Also, I feel like i cn't tell her anything, because she'll think it's stupid, or go and tell the whole world. Joan also says that I treat my other friend better, but i think it's really hard to treat them equally when my OTHER friend is actually a FRIEND... I don't know what to do, I think i still want to be her friend, but I just don't know what to do... can you help?"*

Extract 11

SA : *"I don't know what 2 do I really fancy a boi and he has asked me out but uc mi 2 best friends have a major crush on him and wll probably by enemys wiv me if I do go out wiv him HELP"*

SB : *"Ooooh....tough one :S Get your friends to fancy someone different and then go out with this guy? :D xxxx"*

SC : *"I suggest that you talk to them...n if their actual good friends they will be happy that at least one them go a chance with him, hope that helps"*

SD : *"DO NOT GO OUT WITH HIM! That is totally a recipe for disaster! You all need to forget about him before he comes between your friendship. Boys come and go but you guys will always have each other xxx"SE: "...At the end of the day, it's whatever makes you happy that counts, but remember this: boyfriends come and go, but your friends are anything but temporary!! Hope this helps you!!!:):):)*
*Luv *Real Name**

This solicitation of advice by an individual, as in the case of SA requesting advice on her dilemma regarding a potential boyfriend and her friends is typical of the discourse in children's use of cyberspace to solicit advice. SA is able to explicitly request 'HELP' with her dilemma and this is immediately responded to by four separate offers of advice. Frequently there is no response or feedback from those requesting advice, yet the same individuals continue to provide support and guidance to others whilst making few requests for updates on success in the implementation of advice. An exception is provided below in Extract 12 where BA responds several times with updates and finally thanks her cyberspace peers for their help.

Extract 12

BA : *"heres my problem.....i wonna have .popular gurl friends....i'm already friends with pop.ular boys but i cant only hang out with them bcuz i will be the only gurl in the group...and then popular gurls dont really like me bcuz i have all the guyz to myself...and u know its not easy to jus become friends with them...oh yea! and there are 2 popular gurls at my volleyball team...how should i start talkin to them... bcuz i dont wonna say anythin stupid so they wont think i'm a geek....plz help me somebody!!!!!!!!1*
*luv, *Real Name*"*

BB : *"just be yor self and let ppl like u for who u r"*

BE : *"Talk to people, get used to them, ask for their phone number so you can talk to them more. Say you like them, and wanna be their friends. Thats it. Oh, and, uh, be your self."*

BA : *"ok today i talked to one of the gurls and they were kinda sweet but how can i make them b friend with me i'm so confused HELP!!!!!!!!!!!!!!!!!!!!!! PLZ"*

BF : *"am popular. plz be yourself. talk to popular kids if you want to. popular kids will listen to you. probably not. just be yourself."*

BA : *"does anybody knows how to make them laugh...or any kind of jokes somethin that is not geeky kinda of way like i really want them to like me and i really wonna b friends with those gurls they are really sweet and i can go shopin with them and jus hang out with them at lunch! but when me n my friend tryed to come to their group at lunch they dont reeally talk to us they jus say hey...and when we say wats up (and we're tryin to get into the curcle) it jus doesnt work ...they just start talkin among each other and its really akward so me n my friend have to leave, how can i make them talk to me?"*

BH : *"im popular ill be your friend"*

BN : *"just introduse yourself and start sayn nice stuff bout them. trust me it alllllways works. butttttt if they don't get it ask them to be your friend..."*

BO : *"im not popular but im friends with popular ppl...last year i wasnt...i was a total outcast, but now the boys and the grls r nice to me...all u hav o do is be nice bak."*

BA : *"Ya thanks but i guess they get gellus becuase i gng with only 1 girl in our group and know there is only girl that under stands me!!!!!! thanks ppl 4 ur advice"*

Conclusion

As evidenced in the preceding extracts, children consistently demonstrated in their chatroom interaction that cyberspace was a useful forum for soliciting and providing social support, emotional understanding, affection, and advice about a range issues. This was in line with the prediction of the authors and supports the contention that cyberspace may provide an additional social space in which children can enjoy the developmental benefits of peer interaction.

Clearly there are well documented physical and psychological risks of social interaction in cyberspace posed by the diversity, anonymity and social distance of potential interactive material and partners. However, beyond the risk posed by misinformation, pornography and paedophiles there is also understandable suspicion concerning the potential for internet use to displace other more developmentally 'valuable' activities. It is not the aim of the authors to advocate the displacement of face-to-face interaction by cyberspace interaction, but merely to highlight its benefits, in the absence of, or complementary to, face-to-face opportunities. Considering the nature of the current study, it is prudent and academically responsible to acknowledge the potential harmful impact of children's use of cyberspace. The authors suggest that with due care and information, the same diversity, anonymity and social distance that invokes suspicion of cyberspace, creates an unrivalled developmental space for children. It is a space children can use to observe, practice and even create social behaviours with relative freedom from the physical risks of urban life and requires further illumination from this perspective.

Not only is cyberspace providing social support and clearly of worth for children dealing with 'typical' childhood issues, the authors often observed its value in more extreme cases of childhood experience, such as children who were victims of physical and substance abuse and experiencing compromised mental health. It is perhaps in these situations that the anonymity and non-judgemental characteristics of peer interaction in cyberspace are most clearly discernible. The final extract below not only provides evidence of the willingness of children to utilise the cyberspace environment in response to their perceived lack of social capital, but also clearly illustrates the valuable service provided by the prosocial and supportive environment created by the empathy and understanding of children in cyberspace. It is truly their domain, where they adopt responsibility and ownership of situations where many adults might recoil from the intimacy of the disclosure, and are in turn empowered by one another's needs and responsiveness.

Extract 13

MA : *"i hate my life so much...school is okay but all my friendsare in fights except me and im in the middle and its really stressful...im 14 and never had a true boyfriend. i think im fat...im lonley all the time...i got to an all girls schoool so i have no chance of meeting guys...my friends get mad when i hang out with new ppl...my life just sucks pretty much...but im scarecd to kill myself i dontk nwo what to do please help..."*

MB : *"Relax, girl. You need to talk to somebody, right? That somebody will be me. You do have preety stressfull life, so thats why i am here. Consider me somebody you can talk to about anything to everything. Girl, you are lucky now that you know me! Send me a post dude!"*

MD : *"please please dont give up. last year i felt the same way. but im still here. and im glad of it. before you even attempt suicide think about all the people you will affect friends, family, any1 close 2 u. every1 fights and im sure your friends will make up before you know it. im sure they dont relise how much this is affecting you. u will get a boyfriend, you just havnt found the right person yet, but u will soon. please dont give up, (real name) xx"*

MF : *"don't think about dying you can't take your life only god can take it your life will get so much better even if you don't belive so please belive in yourself and always have faith in you, don't you want to go to college and have a good job, and have a wonderful husband and wonderful kids , just imagine your life if you decide to keep it does'nt it sound like fun please don harm yourself , just belive!!" Hope I helped ! BYE, AND TAKE CARE!!!"*

MG : *"Hey i kno exactly how you feel. I had pretty much the same thing happen to me but trust me sucicide is NOT the answer! I promise u the fights will blow over. I go to an all guys school and i kno how u feel about not finding a match for u, but just wait ur turn will come, just talk with other people and still with ur friends and by doing so ull meet other people. Now about the fat issue, i used to think im fat but then i decided to start running and working out and now i love my body! In ur spare time just run or do sit ups and ur body image will go up! Most of all Never even thing of killing ur self! Just think about all the people that will miss u and how ur family will miss u. Hav u ever thought about life after death? exactly it seems to me like theres nothing. No life with ne1. So just hang in there and keep ur self out of the stressful enviorment!If u need to talk im here!"*

ML : *"15/f/hell - look i know wat u r going through im suicidle tried 2 to kill my self talk to me i can really help u"*

MA : *"thanks you guys, really i dont think ide every consider suicide b.c i had a friend who cut herslef and that was a mess. i think i was just having a really bad day b.c i cant control my life but i want to. i guess i just have to sit tight and wait things out. THANK YOU SO MUCH!!!!!! I LOVE ALL OF YOU"*

REFERENCES

Bargh, J.A., Fitzsimons, G.M. and McKenna, K.Y.A. 2002. "Can You See the Real Me? Activation and Expression of the "True Self" on the Internet", *Journal of Social Issues*, 58(1): 33-48.

Brennan, P.F., Moore, S.M. and Smyth, K.A. 1992. "Alzheimer's disease caregivers' uses of a computer network". *Western Journal of Nursing Research*, 14: 662-673.

Brown, B.B. (1999). "Measuring the peer environment of American adolescents", (pp. 59-90) in S.L. Friedman and T.D. Wachs (eds.), *Assessment of the environment across the life span*. Washington, DC: American Psychological Association.

Bruckman, A. 1992. *Identity workshop: Emergent social and psychological phenomena in textbased virtual reality*. Unpublished manuscript. MIT Media

Laboratory. Cambridge, MA. Available via anonymous ftp from media.mit.edu in pub/MediaMOO/Papers/identity-workshop.

Bukowski, W.M. and Hoza, B. 1989. "Popularity and friendship: issues in theory, measurement, and outcome", (pp. 15-45) in T.J. Berndt & G.W. Ladd (eds.), *Peer relationships in child development*. New York: Wiley.

Coleman, J. 1987. "Families and schools", *Educational Researcher,* 16(6): 32-38.

Finholt, T., and Sproull, L.S. 1990. "Electronic groups at work", *Organization Science*, 1: 41-64.

Gable, S.L., Gross, E.J., and Juvonen, J. 2002. "Internet Use and Well-Being in Adolescence", *Journal of Social Issues,* 58(1): 75-90.

Hartup, W.W. 1996. "The company they keep: Friendships and their developmental significance", *Child Development,* 67: 1-13.

Hasselbring, T.S. and Williams-Glaser, C.H. 2000. Use of computer technology to help students with special needs. Future of Children 10(2). Online: *www.futureofchildren.org/pubs-info2825/pubs-info.htm?doc_id=69787 (02/2005)*

Kiesler, S.B., Siegal, J. and McGuire, T.W. 1984. "Social psychological aspects of computer-mediated communication", *American Psychologist*, 39: 1123-1134.

Kiesler, S. and Sproull, L. 1992. "Group decision making and communication technology", *Organization Behavior and Human Decision Processes,* 52: 96-123.

Ladd, G. 1990. "Having friends, keeping friends, making friends, and being liked by peers in the classroom: predictors of children's early school adjustment?" *Child Development*, 61: 1081-1100.

Lea, M. and Spears, R. 1995. "Love at first byte? Building personal relationships over computer networks", (pp. 197-233) in J.T. Wood & S. Duck (eds.), *Understudied relationships: Off the beaten track.* Newbury Park, CA: Sage.

Livingston, S., Bober, M. and Helsper, E. 2004. "Active participation or just more information?: young people's take up of opportunities to act and interact on the Internet", A research report from the UK Children Go Online project, consulted 30 January, 2005. http://www.children-go-online.net

McCormick, N.B., & McCormick, J.W. 1992. "Computer friends and foes: Content of undergraduates' electronic mail", *Computers in Human Behavior*, 8: 379-405.

McKenna, K.Y.A., & Bargh, J.A. 2000. "Plan 9 from cyberspace: The implications of the Internet for personality and social psychology", *Personality and Social Psychology Review*, 4: 57-75.

Merchant, G. 2001. "Teenagers in cyberspace: language use and language change in internet chatrooms", *Journal of Research in Reading,* 24(3): 293-306.

Myers, D. 1987. "Anonymity is part of the magic: Individual manipulation of computer-mediated communication contexts", *Qualitative Sociology,* 10: 251-266.

Parker, J. and Gottman, J.M. 1989. "Social and emotional development in relational context: Friendship interaction from early childhood to adolescence", (pp. 95-132) in T.J. Berndt & G.W. Ladd (eds.), *Peer relationships in child development*. New York: Wiley.

Parks, M.R. 1995. "Webs of influence in interpersonal relationships", (pp. 155-178) in C.R. Berger & M.E. Burgoon (eds.), *Communication and social influence processes*. East Lansing: Michigan State University Press.

Parks, M.R. 1996. "Making Friends in Cyberspace", *Journal of Communication*, 46(1): 80-98.

Pew Internet and American Life Project. 2001. Teenage life online: The rise of the instant message generation and the Internet's impact on friendships and family relationships [Online, retrieved July 19, 2001]. Available: *http://www.pewinternet.org/reports/pdfs/PIP_Teens_Report.pdf*

Reis, H.T. & Gable, S.L. 2000. "Event-sampling and other methods for studying everyday experience", (pp. 190-222) in H.T. Reis & C.M. Judd (eds.), *Handbook of research methods in social and personality psychology*. New York: Cambridge University Press.

Reis, H.T. & Shaver, P. 1988. "Intimacy as an interpersonal process", (pp. 367-389) in S. Duck (ed.), *Handbook of personal relationships*. Chichester, UK: Wiley.

Reis, H.T. & Sheldon, K.M., Gable, S.L., Roscoe, J. and Ryan, R.M. 2000. "Daily well-being: The role of autonomy, competence, and relatedness", *Personality and Social Psychology Bulletin*, 26: 419-435.

Rice, R.E. 1987. "Computer-mediated communication and organizational innovation", *Journal of Communication*, 37: 65-94.

Rice, R.E. & Love, G. 1987. "Electronic emotion: Socio-emotional content in a computer mediated communication network", *Communication Research*, 14: 85-108.

Rideout, V.J., Foehr, D.F. and Roberts, U.G. 1999. *Kids and media at the new millennium: Executive summary*. Menlo Park, CA: Kaiser Family Foundation.

Short, J., Williams, E. and Christie, B. 1976. *The social psychology of telecommunication*. London: John Wiley.

Strauss, A.L. and Corbin, J. 1990. *Basics of qualitative research: Grounded theory procedures and techniques*. Newbury Park, CA: Sage Publications.

Sproull, L. and Kiesler, S. 1991. *Connections: New ways of working in the networked organization*. Cambridge, MA: MIT Press.

Suler, J. 1998. The Basic Psychological Features of Cyberspace. http://www.rider.edu/users/suler/psycyber/basicfeat.html (01/2005)

Turow, J. 1999, May 4. The Internet and the family: The view from the family, the view from the press [Online, retrieved March 2005]. The Annenberg Public Policy Center of the University of Pennsylvania. *http://www.appcpenn.org/internet/family/rep27.pdf*

Werner, E.E. 1982. "Resilient children", *Young Children*, 10: 68-72.

Wilkins, H. 1991. "Computer talk: Long-distance conversations by computer", *Written Communication*, 8: 56-78.

INDEX